WE17 JAR

The Atlas of Musculo-skeletal Anatomy

Chris Jarmey

Lotus Publishing
Chichester, England

and

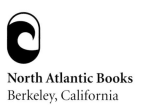

North Atlantic Books
Berkeley, California

First published in 2004 by
Lotus Publishing
9 Roman Way, Chichester, PO19 3QN and
North Atlantic Books
P O Box 12327
Berkeley, California 94712

All Drawings Amanda Williams
Text Design Tracey Shooter
Cover Design Jim Wilkie
Printed and Bound in Singapore by Tien Wah Press

The Atlas of Musculo-skeletal Anatomy is sponsored by the Society for the Study of Native Arts and Sciences, a nonprofit educational corporation whose goals are to develop an educational and crosscultural perspective linking various scientific, social, and artistic fields; to nurture a holistic view of arts, sciences, humanities, and healing; and to publish and distribute literature on the relationship of mind, body, and nature.

British Library Cataloguing in Publication Data
A CIP record for this book is available from the British Library
ISBN 0 9543188 3 8 (Lotus Publishing)
ISBN 1 55643 529 0 (North Atlantic Books)

Library of Congress Cataloging-in-Publication Data

Jarmey, Chris.
 The atlas of musculo-skeletal anatomy / Chris Jarmey.
  p. ; cm.
 Includes bibliographical references and index.
 ISBN 1-55643-529-0 (cloth)
1. Musculoskeletal system--Anatomy--Atlases.
 [DNLM: 1. Musculo-skeletal System--Atlases. WE 17 J37a 2004] I. Title.
 QM100.J37 2004
 611.7--dc22
                  2004005196

Contents

11 Muscles of the Shoulder and Arm

12 Muscles of the Forearm and Hand

13 Muscles of the Hip and Thigh

14 Muscles of the Leg and Foot

About this Book

The Atlas of Musculo-skeletal Anatomy is a comprehensive textbook for students and practitioners of physical therapy, physical training and other disciplines, who require thorough and lucid information about the anatomical structures involved in voluntary physical movement. It is intended that the visual clarity of the illustrations will make it easy to interpret this information at a glance. Also, thorough research and cross referencing of material from an extensive range of authoritative sources has hopefully resulted in a book that can claim an unusual degree of exactitude of information. The effort to achieve this was time well spent, because muscle attachments and actions, nerve supply and arterial supply are areas where minor contradictions can exist between reference material. Such inconsistencies may confuse those who require certainty in such things. What most people want from a reference book is an assurance of accuracy. I believe this book delivers that precision.

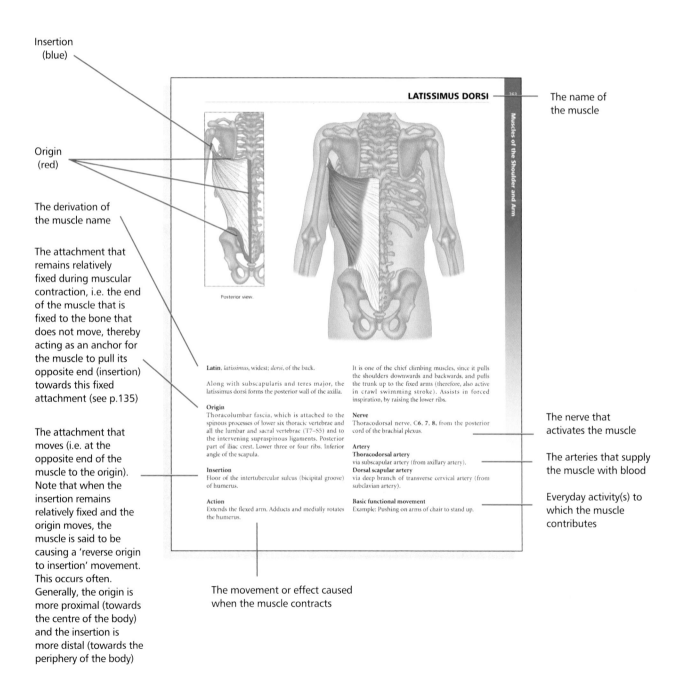

Insertion (blue)

Origin (red)

The derivation of the muscle name

The attachment that remains relatively fixed during muscular contraction, i.e. the end of the muscle that is fixed to the bone that does not move, thereby acting as an anchor for the muscle to pull its opposite end (insertion) towards this fixed attachment (see p.135)

The attachment that moves (i.e. at the opposite end of the muscle to the origin). Note that when the insertion remains relatively fixed and the origin moves, the muscle is said to be causing a 'reverse origin to insertion' movement. This occurs often. Generally, the origin is more proximal (towards the centre of the body) and the insertion is more distal (towards the periphery of the body)

The name of the muscle

The nerve that activates the muscle

The arteries that supply the muscle with blood

Everyday activity(s) to which the muscle contributes

The movement or effect caused when the muscle contracts

LATISSIMUS DORSI

Muscles of the Shoulder and Arm

Posterior view.

Latin. *latissimus*, widest; *dorsi*, of the back.

Along with subscapularis and teres major, the latissimus dorsi forms the posterior wall of the axilla.

Origin
Thoracolumbar fascia, which is attached to the spinous processes of lower six thoracic vertebrae and all the lumbar and sacral vertebrae (T7–S5) and to the intervening supraspinous ligaments. Posterior part of iliac crest. Lower three or four ribs. Inferior angle of the scapula.

Insertion
Floor of the intertubercular sulcus (bicipital groove) of humerus.

Action
Extends the flexed arm. Adducts and medially rotates the humerus.

It is one of the chief climbing muscles, since it pulls the shoulders downwards and backwards, and pulls the trunk up to the fixed arms (therefore, also active in crawl swimming stroke). Assists in forced inspiration, by raising the lower ribs.

Nerve
Thoracodorsal nerve, C6, 7, 8, from the posterior cord of the brachial plexus.

Artery
Thoracodorsal artery
via subscapular artery (from axillary artery).
Dorsal scapular artery
via deep branch of transverse cervical artery (from subclavian artery).

Basic functional movement
Example: Pushing on arms of chair to stand up.

A Note About Peripheral Nerve Supply

The nervous system comprises:

- The central nervous system (i.e. the brain and spinal cord).
- The peripheral nervous system (including the autonomic nervous system, i.e. all neural structures outside the brain and spinal cord).

The peripheral nervous system consists of 12 pairs of cranial nerves and 31 pairs of spinal nerves (with their subsequent branches). The spinal nerves are numbered according to the level of the spinal cord from which they arise (the level is known as the spinal segment).

The relevant peripheral nerve supply is listed with each muscle presented in this book, for those who need to know. However, information about the spinal segment* from which the nerve fibres emanate often differs between the various sources. This is because it is extremely difficult for anatomists to trace the route of an individual nerve fibre through the intertwining maze of other nerve fibres as it passes through its plexus (plexus = a network of nerves: from the Latin word meaning 'braid'). Therefore, such information has been derived mainly from empirical clinical observation, rather than through dissection of the body.

In order to give the most accurate information possible, I have duplicated the method devised by Florence Peterson Kendall and Elizabeth Kendall McCreary (see resources: Muscles Testing and Function). Kendall & McCreary integrated information from six well-known anatomy reference texts; namely, those written by: Cunningham, deJon, Foerster & Bumke, Gray, Haymaker & Woodhall, and Spalteholz. Following the same procedure, and then cross-matching the results with those of Kendall & McCreary, the following system of emphasising the most important nerve roots for each muscle has been adopted in this book.

Let us take the supinator muscle as our example, which is supplied by the deep radial nerve, C5, **6**, (7). The relevant spinal segment is indicated by the letter [C] and the numbers [5, **6**, (7)]. Bold numbers [e.g. **6**] indicate that most (at least five) of the sources agree. Numbers that are not bold [e.g. 5] reflect agreement by three of four sources. Numbers not in bold and in parenthesis [e.g. (7)] reflect agreement by two sources only, or if more than two sources specifically regarded it as a very minimal supply. If a spinal segment was mentioned by only one source, it was disregarded. Hence, bold type indicates the major innervation; not bold indicates the minor innervation; and numbers in parenthesis suggest possible or infrequent innervation.

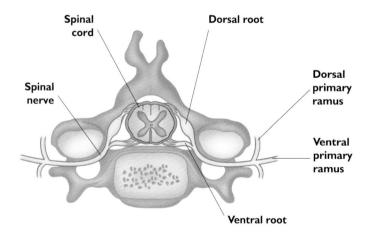

Figure 1a: A spinal segment, showing the nerve roots combining to form a spinal nerve, which then divides into ventral and dorsal rami.

A spinal segment is the part of the spinal cord that gives rise to each pair of spinal nerves (a pair consists of one nerve for the left side and one for the right side of the body). Each spinal nerve contains motor and sensory fibres. Soon after the spinal nerve exits through the foramen (the opening between adjacent vertebrae), it divides into a dorsal primary ramus (directed posteriorly) and a ventral primary ramus (directed laterally or anteriorly). Fibres from the dorsal rami innervate the skin and extensor muscles of the neck and trunk. The ventral rami supply the limbs, plus the sides and front of the trunk.

A Note About Blood Supply to Muscles

When researching the arterial blood supply to each muscle in the body, it became clear that this information is hard to come by if one is looking for clarity and consistency. Different sources sometimes disagree, especially regarding many of the smaller, deeper muscles. Other apparent contradictions merely reflect the emphasis on a different part of the arterial 'chain' through which the blood travels to get to a muscle. For example, some sources will give one of the arteries supplying the rectus abdominis muscle as the superior epigastric artery, while another source will credit the blood supply to the internal thoracic artery. Because the superior epigastric artery is a branch of the internal thoracic artery, this simply indicates that one source is expressing more detail in their description than the other.

In this book I have labelled the arteries immediately supplying each muscle, but have additionally mentioned the arteries 'upstream' that feed into them. Therefore, the blood vessel that actually connects with the muscle is given first and in bold, often with the blood vessel immediately upstream written in bold on the same line. The major artery that is the source of that blood vessel is then given in plain text and in parenthesis. Where applicable, I have mentioned an intermediary connecting artery if the 'chain' of arteries to a muscle is extensive. For example, in the case of the iliacus muscle, the blood supply is expressed thus:

Iliolumbar branch of the internal iliac artery
via common iliac artery (from abdominal aorta).

So, if the iliolumbar branch artery is analogous to an irrigation channel that branches off from the internal iliac artery, and the internal iliac artery is itself fed from the more central abdominal aorta via the common iliac artery, we have a comprehensive overview of the route taken by the blood to reach its target.

Where a muscle is clearly supplied by a certain artery, but *may* also be supplied by a secondary artery (either because it applies in some people but not all, or because many but not all authorities agree that it does), I have added the potential secondary supply in plain, as shown below:

Inferior gluteal artery
via internal iliac artery (a branch of the common iliac artery from abdominal aorta), plus can also be supplied by medial circumflex arteries (from deep femoral artery).

Where there is more than one blood supply of more or less equal importance (as is the case with the diaphragm) it is shown in the following format:

Musculophrenic artery
via internal thoracic artery (from subclavian artery).
Superior phrenic artery
(from thoracic aorta).
Inferior phrenic artery
(from abdominal aorta).

Overleaf are basic illustrations of peripheral nerve distribution; arterial and venous distribution. As this book is primarily about musculo-skeletal anatomy (muscles, bones and joints), greater detail in these areas was considered unnecessary.

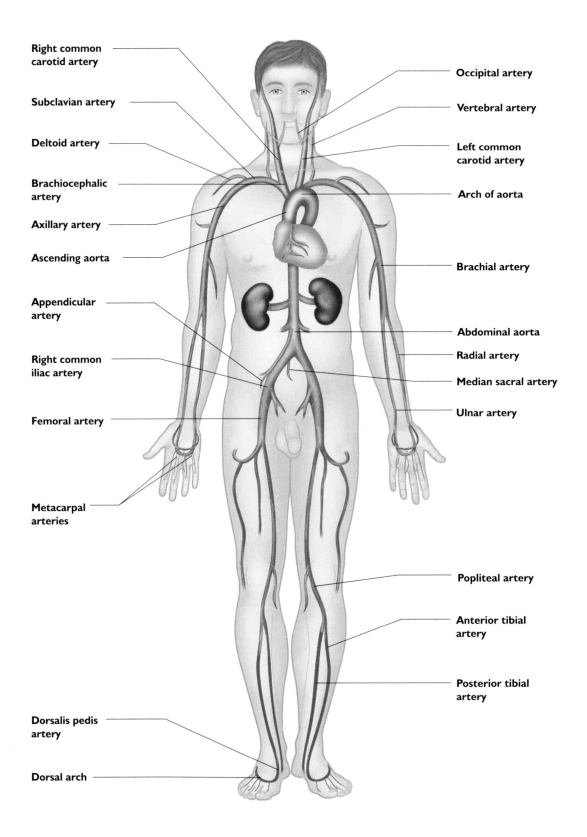

Right common carotid artery

Subclavian artery

Deltoid artery

Brachiocephalic artery

Axillary artery

Ascending aorta

Appendicular artery

Right common iliac artery

Femoral artery

Metacarpal arteries

Dorsalis pedis artery

Dorsal arch

Occipital artery

Vertebral artery

Left common carotid artery

Arch of aorta

Brachial artery

Abdominal aorta

Radial artery

Median sacral artery

Ulnar artery

Popliteal artery

Anterior tibial artery

Posterior tibial artery

Figure 1b: A general overview of major arteries and branches.

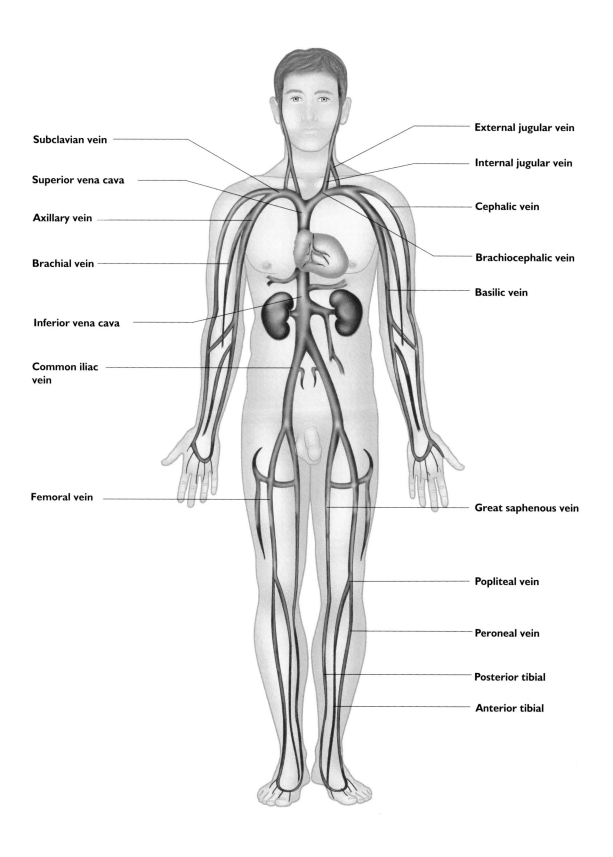

Subclavian vein

Superior vena cava

Axillary vein

Brachial vein

Inferior vena cava

Common iliac vein

Femoral vein

External jugular vein

Internal jugular vein

Cephalic vein

Brachiocephalic vein

Basilic vein

Great saphenous vein

Popliteal vein

Peroneal vein

Posterior tibial

Anterior tibial

Figure 1c: A general overview of major veins and branches.

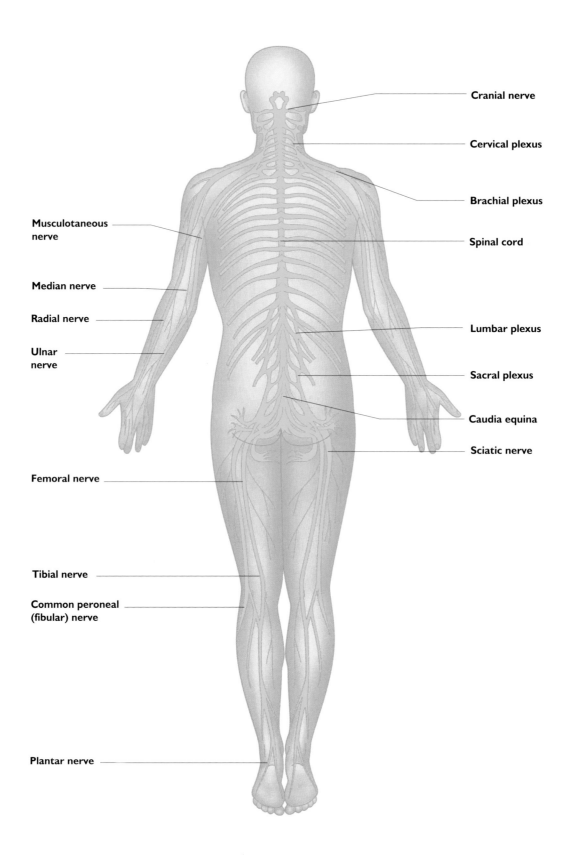

Cranial nerve

Cervical plexus

Brachial plexus

Musculotaneous nerve

Spinal cord

Median nerve

Radial nerve

Lumbar plexus

Ulnar nerve

Sacral plexus

Caudia equina

Sciatic nerve

Femoral nerve

Tibial nerve

Common peroneal (fibular) nerve

Plantar nerve

Figure 1d: A general overview of major peripheral nerves.

Anatomical Orientation

1

Anatomical Directions

To describe the relative position of body parts and their movements, it is essential to have a universally accepted initial reference position. The standard body position known as the anatomical position serves as this reference. The *anatomical position* is simply the upright standing position with arms hanging by the sides, palms facing forwards (*see* figure 2). Most directional terminology used refers to the body *as if* it were in the anatomical position, regardless of its actual position. Note also that the terms 'left' or 'right' refer to the sides of the object or person being viewed, and not those of the reader.

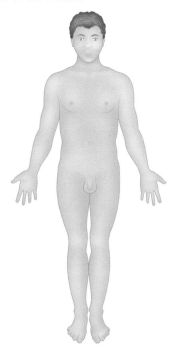

Figure 2: **Anterior.**
In front of; toward or at the front of the body.

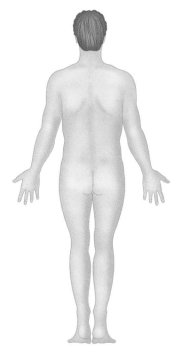

Figure 3: **Posterior.**
Behind; toward or at the backside of the body.

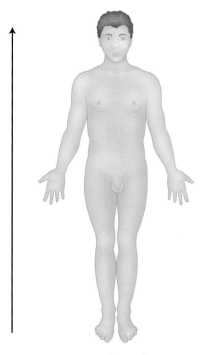

Figure 4: **Superior.**
Above; toward the head or upper part
of the structure or the body.

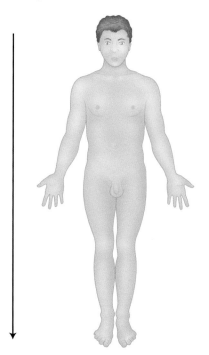

Figure 5: **Inferior.**
Below; away from the head or toward the
lower part of a structure or the body.

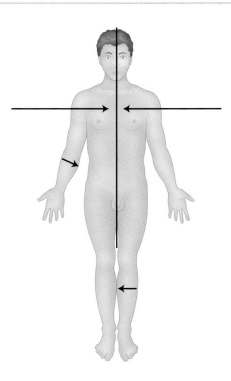

Figure 6: **Medial.**
(from *medius* in Latin, meaning middle)
Toward or at the midline of the body;
on the inner side of a limb.

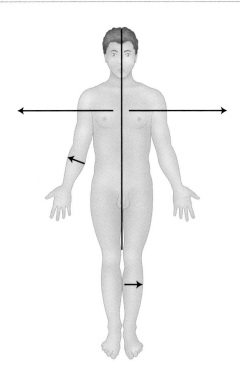

Figure 7: **Lateral.**
(from *latus* in Latin, meaning side)
Away from the midline of the body;
on the outer side of the body or a limb.

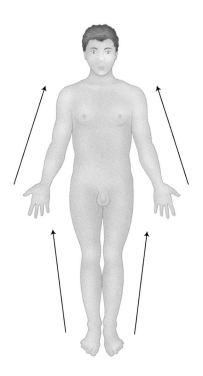

Figure 8: **Proximal.**
(from *proximus* in Latin, meaning next to)
Closer to the centre of the body (the navel), or to the point
of attachment of a limb to the body torso.

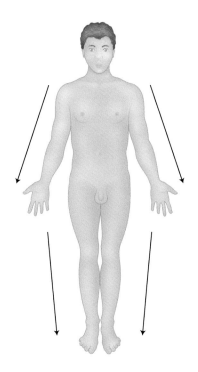

Figure 9: **Distal.**
(from *distans* in Latin, meaning distant)
Farther from the centre of the body, or from the point
of attachment of a limb to the torso.

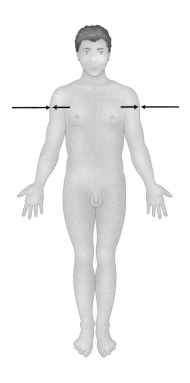

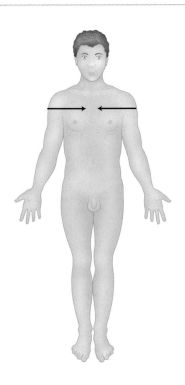

Figure 10: **Superficial.**
Toward or at the body surface.

Figure 11: **Deep.**
Farther away from the body surface;
more internal.

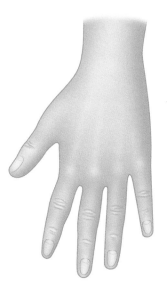

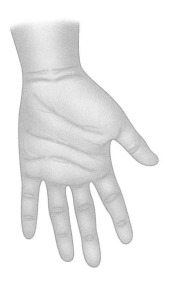

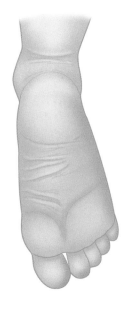

Figure 12: **Dorsum.**
The posterior surface of something,
e.g. the back of the hand;
the top of the foot.

Figure 13: **Palmar.**
The anterior surface of the hand,
i.e. the palm.

Figure 14: **Plantar.**
The sole of the foot.

Regional Areas

The two primary divisions of the body are its *axial* part, consisting of the head, neck and trunk, and its *appendicular* parts, consisting of the limbs that are attached to the axis of the body. Figure 15 shows the terms used to indicate specific body areas. Terms enclosed within brackets refer to the lay term for the area.

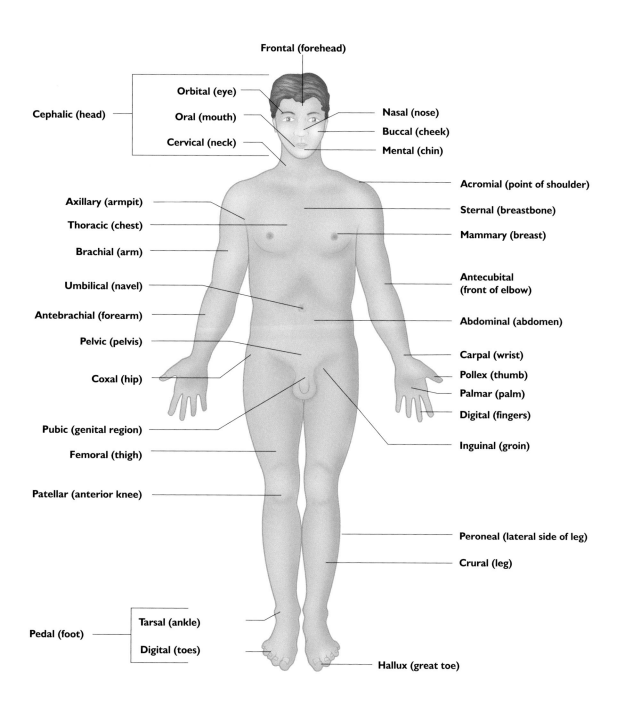

Figure 15: Terms used to indicate specific body areas;
a) anterior view.

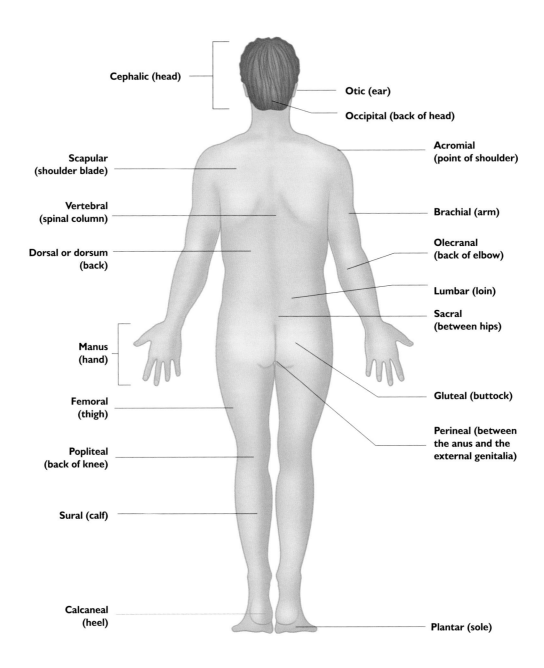

Cephalic (head)

Otic (ear)

Occipital (back of head)

Acromial
(point of shoulder)

Scapular
(shoulder blade)

Vertebral
(spinal column)

Brachial (arm)

Dorsal or dorsum
(back)

Olecranal
(back of elbow)

Lumbar (loin)

Sacral
(between hips)

Manus
(hand)

Femoral
(thigh)

Gluteal (buttock)

Perineal (between
the anus and the
external genitalia)

Popliteal
(back of knee)

Sural (calf)

Calcaneal
(heel)

Plantar (sole)

Figure 15: Terms used to indicate specific body areas;
b) posterior view.

Planes of the Body

Planes refer to two-dimensional sections through the body, to give a view of the body or body part, as if it has been cut through an imaginary line.

- The sagittal planes cut vertically through the body from anterior to posterior, dividing the body into right and left halves. The illustration shows the mid-sagittal plane.
- The frontal (coronal) planes pass vertically through the body, dividing the body into anterior and posterior sections, and lies at right angles to the sagittal plane.
- The transverse planes are horizontal cross sections, dividing the body into upper (superior) and lower (inferior) sections, and lie at right angles to the other two planes. Figure 16 illustrates the most frequently used planes.

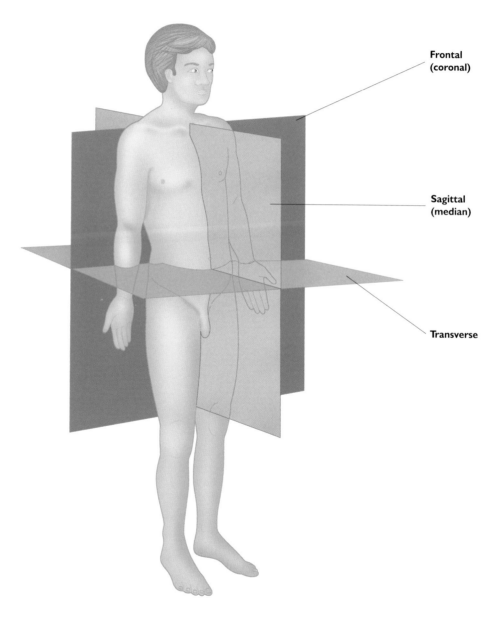

Frontal
(coronal)

Sagittal
(median)

Transverse

Figure 16: Planes of the body.

Anatomical Movements

The direction that body parts move is described in relation to the foetal (fetal) position. Moving into the foetal position results from flexion of all the limbs. Straightening out of the foetal position results from extension of all the limbs.

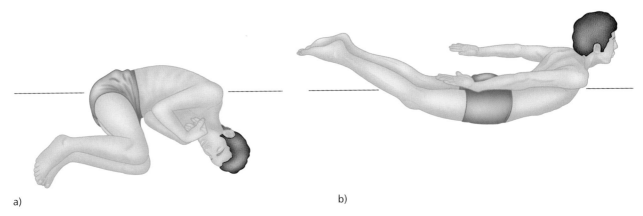

a)   b)

Figure 17: a) Flexion into the foetal position; b) extension out of the foetal position.

Main Movements

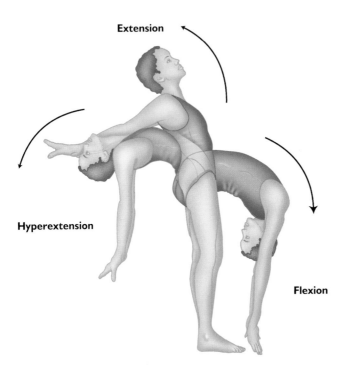

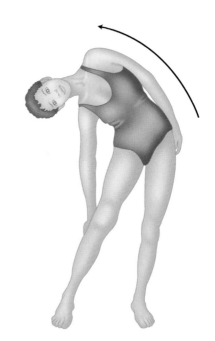

Figure 18:
Flexion: Bending to decrease the angle between bones at a joint. From the anatomical position, flexion is usually forward, except at the knee joint where it is backward. The way to remember this is that flexion is always toward the foetal position. **Extension:** To straighten or bend backward away from the foetal position.
Hyperextension: means to extend the limb beyond its normal range.

Figure 19: **Lateral flexion.**
To bend the torso or head laterally (sideways) in the frontal (coronal) plane.

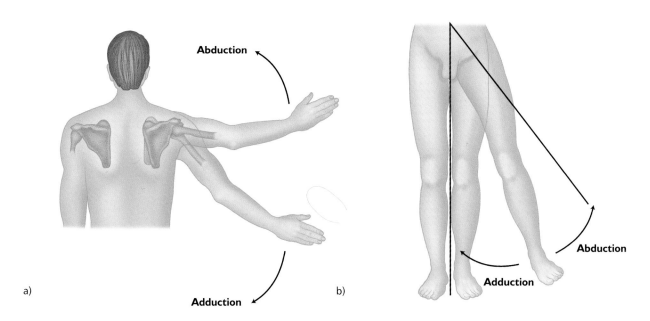

Figure 20a and b:
Abduction: Movement of a bone away from the midline of the body or the midline of a limb.
Adduction: Movement of a bone towards the midline of the body or the midline of a limb.

NOTE: for abduction of the arm to continue above the height of the shoulder (elevation through abduction, *see* p.25), the scapula must rotate on its axis to turn the glenoid cavity upwards (*see* figure 28b).

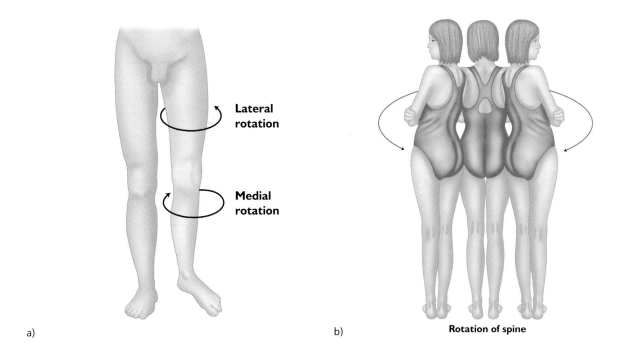

Figure 21:
Rotation: Movement of a bone or the trunk around its own longitudinal axis.
Medial rotation: to turn in towards the midline.
Lateral rotation: to turn out, away from the midline.

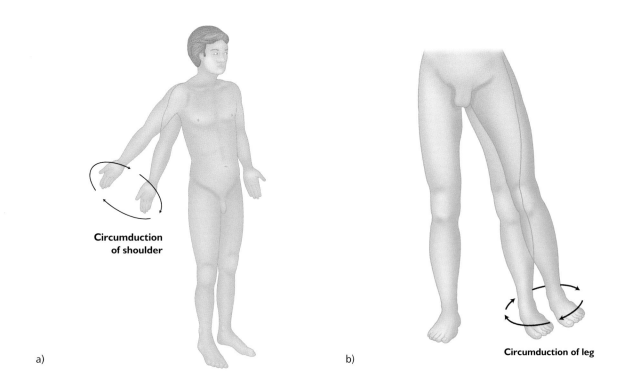

a)   b)

Figure 22: **Circumduction.**
Movement in which the distal end of a bone moves in a circle, while the proximal end remains stable;
the movement combines flexion, abduction, extension, and adduction.

Other Movements

Movements in this section are those that occur only at specific joints or parts of the body; usually involving more than one joint.

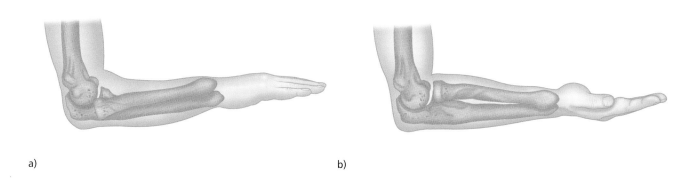

a)   b)

Figure 23a: **Pronation.**
To turn the palm of the hand down to face the floor
(if standing with elbow bent 90°, or if lying flat on the floor),
or away from the anatomical and foetal positions.

Figure 23b: **Supination.**
To turn the palm of the hand up to face the ceiling
(if standing with elbow bent 90°, or if lying flat on the floor),
or toward the anatomical and foetal positions.

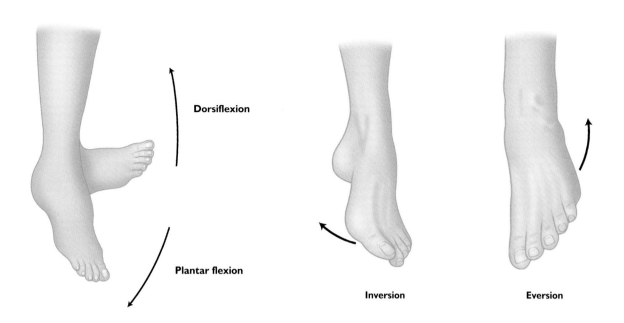

Figure 24: **Plantar flexion:** To point the toes down towards the ground. **Dorsiflexion**: To point the toe towards the sky.

Figure 25: **Inversion:** To turn the sole of the foot inward, so that the soles would face towards each other. **Eversion:** To turn the sole of the foot outward, so that the soles would face away from each other.

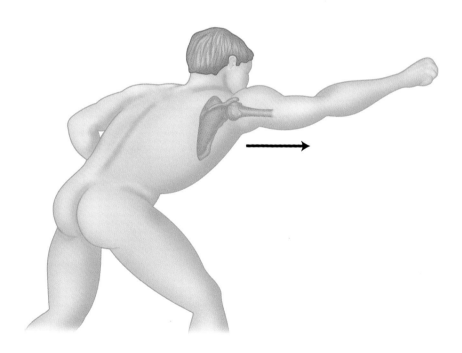

Figure 26: **Protraction.**
Movement forwards in the transverse plane.
For example, protraction of the shoulder girdle, as in rounding the shoulder.

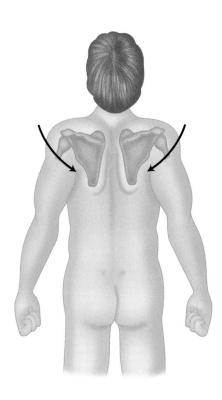

Figure 27: **Retraction.**
Movement backward in the transverse plane,
as in bracing the shoulder girdle back, military style.

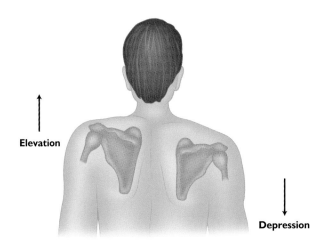

Figure 28a:
Elevation: Movement of a part of the body upwards along the frontal plane.
For example, elevating the scapula by shrugging the shoulders.
Depression: Movement of an elevated part of the body downward to its original position.

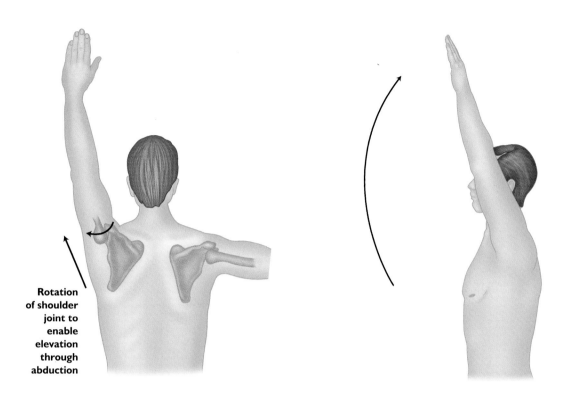

Rotation of shoulder joint to enable elevation through abduction

Figure 28b: Abducting the arm at the shoulder joint, then continuing to raise it above the head in the frontal plane can be referred to as **elevation through abduction.**

Figure 28c: Flexing the arm at the shoulder joint, then continuing to raise it above the head in the sagittal plane can be referred to as **elevation through flexion.**

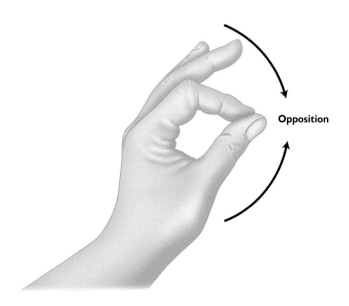

Opposition

Figure 29: **Opposition**.
A movement specific to the saddle joint of the thumb, that enables you to touch your thumb to the tips of the fingers of the same hand.

Tissues

2

Groups of cells having a similar structure and function are called tissues. There are four primary types of tissue, each having a characteristic function:

1. Epithelial

For example, the skin, whose function is to cover and protect the body.

2. Connective

For example, ligaments, tendons, fascia, cartilage, and bone, whose functions are to protect, support, and bind together different parts of the body. It is the most abundant and widely distributed tissue type.

3. Muscle

The function of muscle is to enable movement of the body.

4. Nervous

Nerves control body functions and movement.

The types of tissue that concern us regarding musculo-skeletal anatomy are: *connective tissue* and *muscle tissue*, for which a brief overview is given below:

Connective Tissue

Common Characteristics

1. Variations in blood supply – Most connective tissue is well vascularized (having blood vessels), but tendons and ligaments have a poor blood supply, and cartilage is avascular (having no blood vessels), therefore these structures heal very slowly.

2. Extracellular matrix – Connective tissues are made up of many different types of cells and varying amounts of nonliving substance that surrounds the cells. This substance is called *extracellular matrix*. This matrix is produced by connective tissue cells and then secreted to their exterior. Depending on the type of tissue, the matrix may be: liquid, gel-like, semisolid, or very hard.

Because of the matrix, connective tissue is able to bear weight, and to withstand stretching and other abuses, such as abrasion. The matrix contains various types and amounts of fibres, e.g. collagen, elastic, or reticular.

Types of Connective Tissue

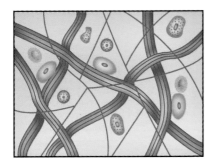

a)

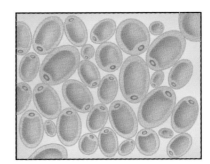

b)

Figure 30: Structure of loose connective tissue; a) areolar, b) adipose.

Loose Connective Tissue

Loose connective tissue has more cells and fewer fibres; thus, it is softer than the other types. Examples are:

1. *Areolar*, a 'packing' tissue, which cushions and protects body organs and holds internal organs together in their proper position;
2. *Adipose*, fat tissue that forms the subcutaneous layer beneath the skin, also called the *hypodermis*, or *superficial fascia*, where it insulates the body and protects against heat and cold.

Dense Regular Connective Tissue

Within dense regular connective tissue, collagen fibres are the predominant element and create a white, flexible tissue with great resistance to pulling forces. Examples are: *ligaments* and *tendons*.

Dense Irregular Connective Tissue

Dense irregular connective tissue has the same structural elements as regular connective tissue. However the bundles of collagen fibres are thicker, are interwoven and are arranged irregularly. Fascia is an example of dense irregular connective tissue.

Figure 31: Structure of dense regular connective tissue.

Figure 32: Structure of dense irregular connective tissue.

Cartilage

Cartilage is tough, but flexible. It has qualities intermediate between dense connective tissue and bone. Cartilage is avascular and devoid of nerve fibres, and therefore heals slowly. Examples are: *hyaline*, *fibrocartilage*, and *elastic*.

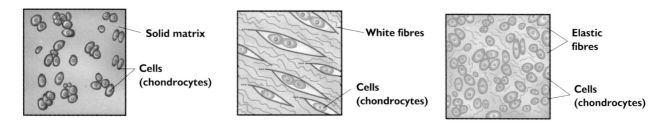

Figure 33: Structure of cartilage; a) hyaline cartilage, b) white fibrocartilage, c) yellow elastic cartilage.

Bone

Bone cells sit in cavities called *lacunae* (sing. lacuna) surrounded by circular layers of a very hard matrix that contains calcium salts and larger amounts of collagen fibres.

Blood

Blood, or vascular tissue, is considered a connective tissue because it consists of blood cells, surrounded by a

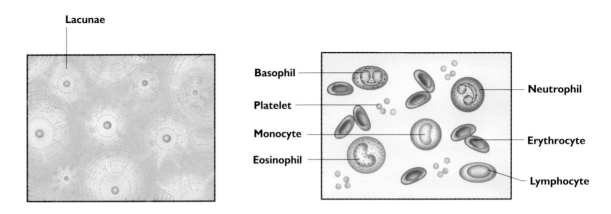

Figure 34: Structure of bone.

Figure 35: Structure of blood.

nonliving fluid matrix called *blood plasma*. The 'fibres' of blood are soluble protein molecules that become visible only during blood clotting. Blood is not a typical connective tissue; it is the transport vehicle for the cardiovascular system, carrying nutrients, wastes, respiratory gases, etc. throughout the body.

Muscle Tissue

Muscle is composed of 75% water, 20% protein and 5% mineral salts, glycogen and fat. As this book is designed to focus specifically on musculo-skeletal anatomy, only a brief description and comparison of the different types of muscle tissue is given below. Skeletal muscle is then discussed in more detail (*see* Chapter 7).

Muscle Types and Function

There are three types of muscle tissue: skeletal, cardiac and smooth. All muscle cells have an elongated shape and are therefore referred to as *muscle fibres*.

Smooth / Unstriated / Involuntary Muscle

Smooth muscle cells are usually spindle-shaped and arranged in sheets or layers. Smooth muscles are found in the *viscera*, i.e. stomach, small and large intestines, blood vessels, uterus (i.e. the hollow organs).

Smooth muscle in the blood vessels contracts to move the blood in the arteries. Smooth muscle also squeezes substances through the organs and tracts. They are under *involuntary* control (although some individuals can train their minds to achieve some control over smooth muscle contractions). Contractions are usually gentle and rhythmic, with the obvious exceptions of vomiting and birth contractions.

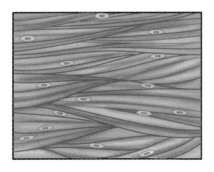

Figure 36: Structure of smooth / unstriated / involuntary muscle.

Cardiac / Striated / Involuntary Muscle

Cardiac muscles are found in the heart only, and exist to pump the heart. They are under *involuntary* control. Structurally, they are made up of branching fibres that are striated in appearance and are separated or interspersed by discs, known as *intercalated discs*.

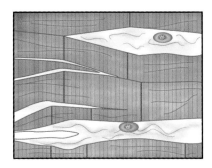

Figure 37: Structure of cardiac / striated / involuntary muscle.

Skeletal / Striated / Voluntary Muscle

Skeletal muscles (also called somatic muscles) attach to, and cover over, the bony skeleton. They are under *voluntary* control. Skeletal muscles fatigue easily, but can be strengthened. They are capable of powerful, rapid contractions, and longer, sustained contractions. Skeletal muscles enable us to perform both feats of strength and controlled, fine movements.

NOTE: As they contract, all muscle types generate heat, and this heat is vitally important in maintaining a normal body temperature. It is estimated that 85% of all body heat is generated by muscle contractions.

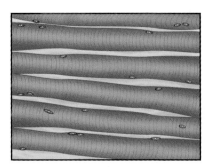

Figure 38: Structure of skeletal / striated / voluntary muscle.

Bone

3

We are born with approximately 350 bones, but gradually they fuse together until by puberty we have only 206 bones. These bones form the supporting structure of the body, and are collectively known as the endoskeleton. (The exoskeleton is well developed in many invertebrates, but exists in humans only as teeth, nails and hair). Fully developed bone is the hardest tissue in the body and is composed of 20% water, 30% to 40% organic matter and 40% to 50% inorganic matter.

Bone Development and Growth

The majority of bone is formed from a foundation of cartilage (*see* below), which becomes calcified and then ossified to form true bone. This process occurs through the following stages:

1. Bone building cells called *osteoblasts* become active during the second or third month of embryonic life.
2. Initially, the osteoblasts manufacture a *matrix* of material between the cells, which is rich in a fibrous protein called *collagen*. This collagen strengthens the tissue. Enzymes then enable calcium compounds to be deposited within the matrix.
3. This intercellular material hardens around the cells, to become *osteocytes*; i.e. living cells that maintain the bone, but do not produce new bone.
4. Other cells, called *osteoclasts*, breakdown, remodel and repair bone; a process that continues throughout life, but which slows down with advancing age. Consequently, the bones of elderly people are weaker and more fragile.

In brief, osteoblasts and osteoclasts are the cells that lay down and break down bone respectively; enabling bones to very slowly adapt in shape and strength according to need.

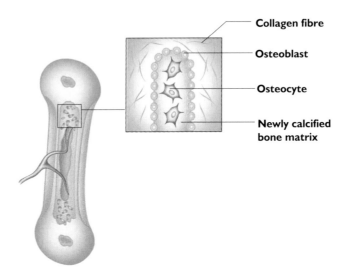

Figure 39: Bone development and growth.

Cartilage

Cartilage (gristle) exists either as a temporary formation that is later replaced by bone, or as a permanent supplementation to bone. However, it is not as hard or as strong as bone.

It consists of living cells called *chondrocytes*, contained within *lacunae* (spaces) and surrounded by a collagen rich intercellular substance. Cartilage is relatively non-vascular (not penetrated by blood vessels) and is mainly nourished by surrounding tissue fluids. There are three main types of cartilage: hyaline cartilage, white fibrocartilage and yellow fibrocartilage.

Hyaline Cartilage

Hyaline cartilage forms the temporary foundation of cartilage from which many bones develop; thereafter existing in relation to bone as:

- The articular cartilage of synovial joints.
- Cartilage plates between separately ossifying areas of bone during growth.
- The xiphoid process of the sternum (which ossifies late or not at all) and the costal cartilages.

Hyaline cartilage also exists in the nasal septum, most cartilages of the larynx and the supporting rings of the trachea and bronchi.

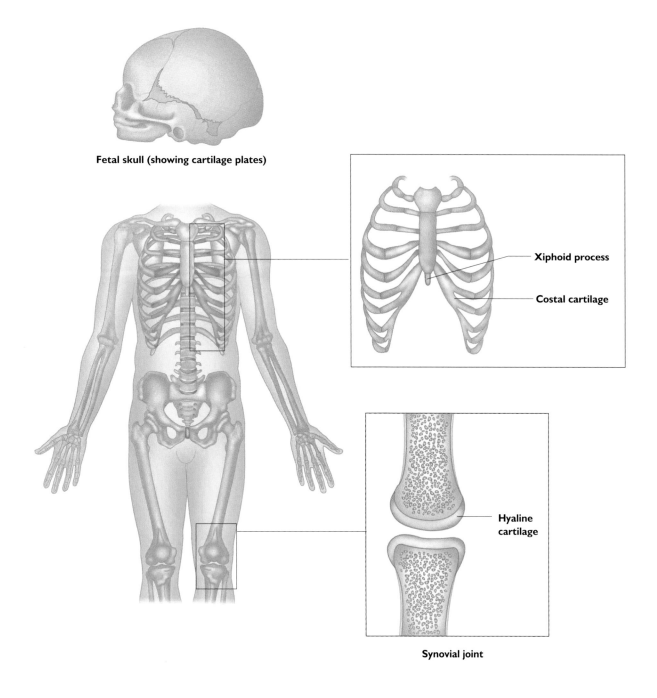

Fetal skull (showing cartilage plates)

Xiphoid process

Costal cartilage

Hyaline cartilage

Synovial joint

Figure 40: Location sites of hyaline cartilage in the body.

White Fibrocartilage

White fibrocartilage contains white fibrous tissue. It has more elasticity and tensile strength than hyaline cartilage. It is found as the:

- Sesamoid cartilages in a few tendons.
- Articular discs in the wrist joint and clavicular joints.
- Rim (labrum) deepening the sockets of the shoulder and hip joints.
- Two semilunar cartilages within each knee joint.
- Intervertebral discs between adjacent surfaces of the vertebral bodies.
- Cartilage plate joining the hipbones at the pubic symphysis.

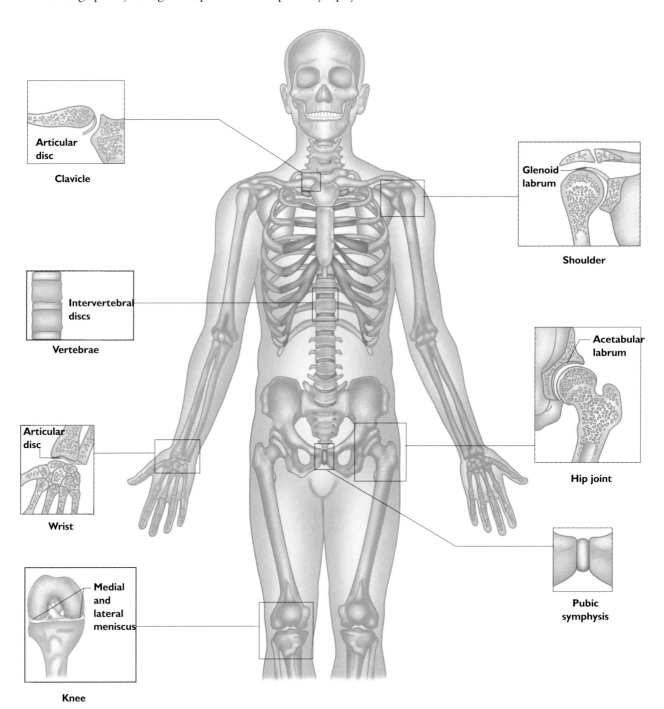

Figure 41: Location sites of white fibrocartilage in the body.

Yellow Fibrocartilage

Yellow fibrocartilage contains yellow elastic fibres. It is found in the external ear, auditory tube of the middle ear, and the epiglottis.

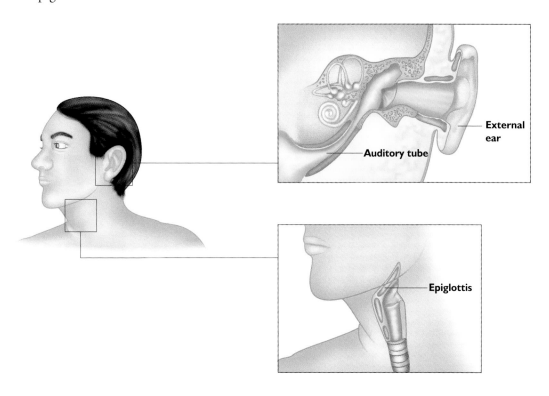

Figure 42: Location sites of yellow fibrocartilage in the body.

Functions of Bones

Support

Our bones provide the hard framework that supports and anchors all the soft organs of the body. Our legs support the body's torso, head, and arms. The ribcage supports the chest wall.

Protection

The bones of the skull protect the brain; the vertebrae surround the spinal cord; the ribcage protects all the vital organs.

Movement

Muscles are attached to the bones by tendons; and they use the bones as levers to move the body and all its parts; the arrangement of the bones and joints determines which movements are possible.

Storage

Fat is stored as 'yellow marrow' in the central cavities of long bones. Within the structure of the bone itself minerals are stored. The most important minerals are calcium and phosphorus, but potassium, sodium, sulphur, magnesium, and copper are also stored. Stored minerals can be released into the bloodstream for distribution to all parts of the body as needed.

Blood Cell Formation

The bulk of blood cell formation occurs within the 'red marrow' cavities of certain bones.

Types of Bone–according to density

Compact Bone

Compact bone is dense, and looks smooth to the naked eye. Through the microscope, compact bone appears as an aggregation of *Haversian systems*, also called *osteons*. Each such system is an elongated cylinder oriented along the long axis of the bone, consisting of a central *Haversian canal* containing blood vessels, lymph vessels and nerves, surrounded by concentric plates of bone called *lamellae*. In other words, each Haversian system is a group of hollow tubes of bone matrix (lamellae), placed one inside the next. Between these lamellae there are spaces (*lacunae*) that contain lymph and osteocytes. The lacunae are linked via hair-like canals called *canaliculi* to the lymph vessels in the Haversian canal, enabling the osteocytes to obtain nourishment from the lymph. This tubular array of lamellae gives great strength to the bone.

Other canals called *perforating*, or *Volkmann's canals*, run at right angles to the long axis of the bone, connecting the blood vessels and nerve supply within the bone to the periosteum (*see* p.41).

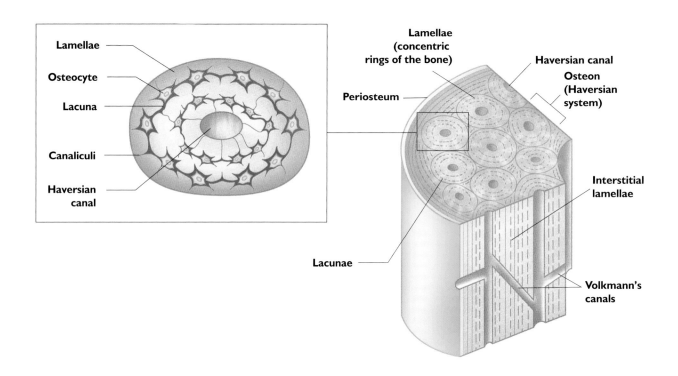

Figure 43: Structure of compact bone.

Spongy Bone (Cancellous Bone)

Spongy bone is composed of small needle-like *trabeculae* (sing. trabecula; literally, '*little beam(s)*') containing irregularly arranged lamellae and osteocytes, interconnected by canaliculi. There are no Haversian systems, but rather, lots of open spaces, that can be thought of as large Haversian canals, giving a honeycombed appearance. These spaces are filled with red or yellow marrow and blood vessels.

This structure forms a dynamic lattice capable of gradual alteration through realignment, in response to stresses of weight, postural change, and muscle tension. Spongy bone is found in the epiphyses of long bones, the bodies of the vertebrae, and other bones without cavities.

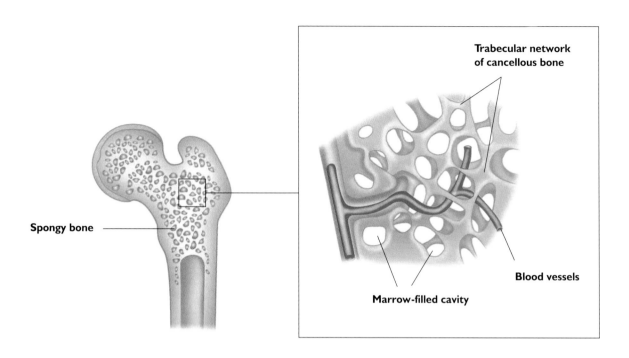

Figure 44: Structure of spongy (cancellous) bone.

Types of Bones–according to shape

Irregular Bones
Irregular bones have complicated shapes; they consist mainly of spongy bone enclosed by thin layers of compact bone. Examples include: some skull bones, the vertebrae, and the hipbones.

Flat Bones
Flat bones are thin, flattened bones, and frequently curved; they have a layer of spongy bone sandwiched between two thin layers of compact bone. Examples include: most of the skull bones, the ribs, and the sternum.

Short Bones
Short bones are generally cube-shaped; consist mostly of *spongy* (*cancellous*) bone. Examples include: the carpal bones in the hand, and tarsal bones in the ankle.

– **sesamoid** (from the Latin, meaning '*shaped like a sesame seed*'): Sesamoid bones are a special type of short bone, that are formed and embedded within a tendon. Examples are: the patella (kneecap) and the pisiform bone at the medial end of the wrist crease.

Long Bones
Long bones are longer than they are wide; they have a shaft with heads at both ends, and consist mostly of compact bone. Examples include: the bones of the limbs, except those of the wrist, hand, ankle and foot (although the bones of the fingers and toes are effectively miniature long bones).

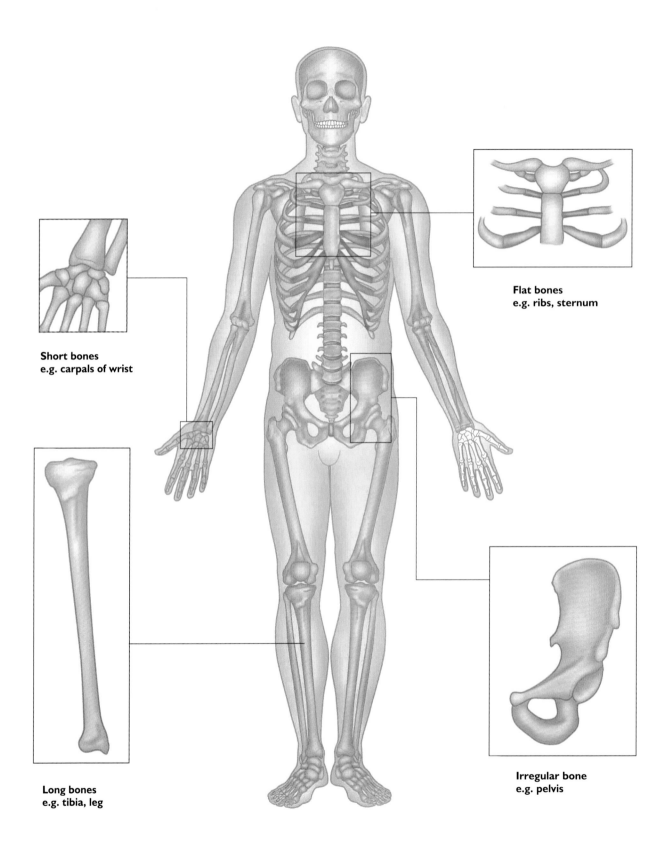

Flat bones
e.g. ribs, sternum

Short bones
e.g. carpals of wrist

Long bones
e.g. tibia, leg

Irregular bone
e.g. pelvis

Figure 45: Bone shapes.

Components of a Long Bone

The transformation of cartilage within a long bone begins at the centre of the shaft. Secondary bone-forming centres develop later on, across the ends of the bones. From these growth centres, the bone continues to grow through childhood and adolescence, finally ceasing in the early twenties; whereupon the growth regions harden.

Diaphysis (from Greek, meaning 'a separation')
The diaphysis is the shaft or central part of a long bone. It has a marrow filled cavity (medullary cavity) surrounded by compact bone. It is formed from one or more primary sites of ossification, and supplied by one or more nutrient arteries.

Epiphysis (from Greek, meaning 'excresence')
The epiphysis is the end of a long bone, or any part of a bone separated from the main body of an immature bone by cartilage. It is formed from a secondary site of ossification. It consists largely of *spongy bone*.

Epiphyseal Line
The epiphyseal line is the remnant of the epiphyseal plate (a flat plate of hyaline cartilage) seen in young, growing bone. It is the site of growth of a long bone. By the end of puberty, long bone growth stops and this plate is completely replaced by bone, leaving just the line to mark its previous location.

Articular Cartilage
Articular cartilage is the only remaining evidence of an adult bone's cartilaginous past. It is located where two bones meet (articulate) within a synovial joint. It is smooth, slippery, porous, malleable, insensitive, and bloodless. It is massaged by movement, which permits absorption of synovial fluid, oxygen, and nutrition.

NOTE: The degenerative process of osteoarthritis (and the latter stages of some forms of rheumatoid arthritis) involves the breakdown of articular cartilage.

Periosteum
The periosteum is a fibrous connective tissue membrane that is vascular and provides a highly sensitive double-layered life support sheath enveloping the outer surface of bone. The outer layer is made of dense irregular connective tissue. The inner layer, which lies directly against the bone surface, mostly comprises the bone-forming *osteoblasts* and the bone-destroying *osteoclasts*.

The periosteum is supplied with nerve fibres, lymphatic vessels, and blood vessels that enter the bone through *nutrient canals*. It is attached to the bone by collagen fibres, known as *Sharpey's fibres*. The periosteum also provides the anchoring point for tendons and ligaments.

Medullary Cavity
The medullary cavity is the cavity of the diaphysis (i.e. the central section of a long bone). It contains marrow: red in the young, turning to yellow in many bones in maturity.

Red Marrow
Red marrow is a red, gelatinous substance composed of red and white blood cells in a variety of developmental forms. The **red marrow cavities** are typically found within the spongy bone of long bones and flat bones. In adults the red marrow, which creates new red blood cells, occurs only in the head of the femur and the head of the humerus, and, much more importantly, in the flat bones such as the sternum and irregular bones, such as the hipbones. These are the sites routinely used for obtaining red marrow samples when problems with the blood-forming tissues are suspected.

Yellow Marrow
Yellow marrow is a fatty connective tissue that no longer produces blood cells.

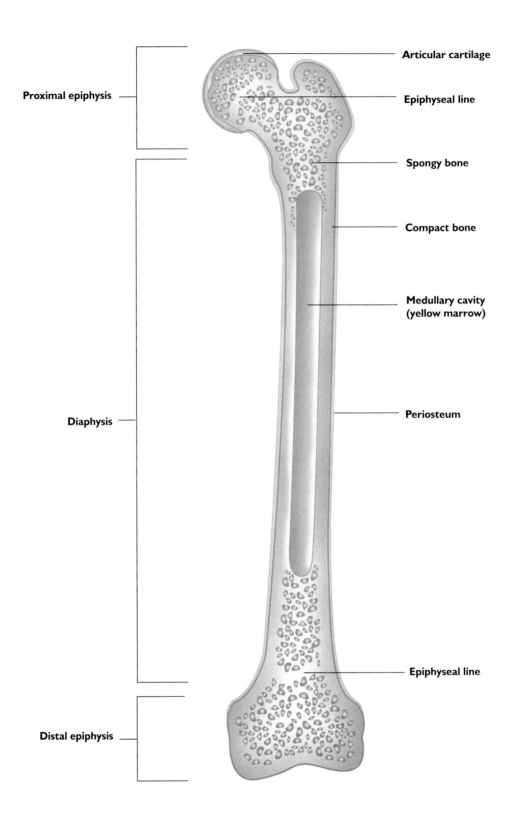

Proximal epiphysis

Articular cartilage

Epiphyseal line

Spongy bone

Compact bone

Medullary cavity
(yellow marrow)

Diaphysis

Periosteum

Epiphyseal line

Distal epiphysis

Figure 46: Components of a long bone.

Bone Markings

Bone markings fall into three broad categories, as given below:

1. Projections on Bones that are the Sites of Muscle and Ligament Attachment

Trochanter
A very large, blunt, irregularly shaped projection. The only example is on the femur.

Tuberosity
A large rounded projection, which may be roughened. The main examples are on the tibia (tibial tuberosities) and the ischium (ischial tuberosities).

Tubercle
A smaller rounded projection, which may be roughened.

Crest
A narrow ridge of bone, usually prominent; notably the iliac crest.

Border
A narrow ridge of bone that separates two surfaces.

Spine or Spinous Process
A sharp, slender, often pointed projection; notably the spinous processes on the vertebrae; and the spines of the scapula or the ilium (anterior superior iliac spine, abbreviated as the ASIS, and the posterior superior iliac spine, abbreviated as the PSIS).

Epicondyle
A raised area on or above a condyle; notably on the humerus at the elbow joint.

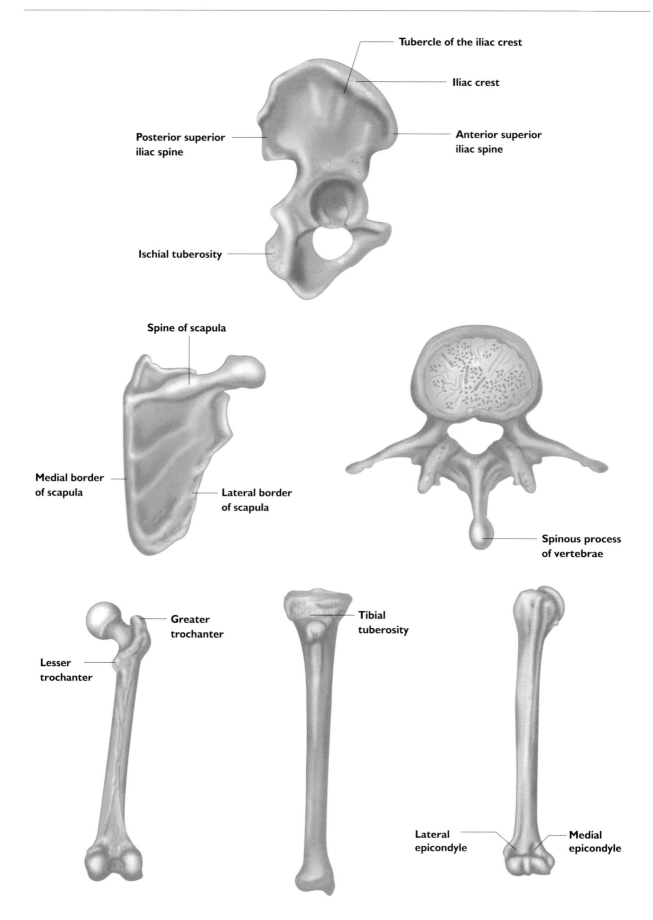

Figure 47: Projections on bones that are the sites of muscle and ligament attachment.

2. Projections on Bones that Help to Form Joints

Head

An expansion which is usually round, located at one end of a bone; e.g. the head of the fibula, which articulates with the tibia just below the knee joint.

Facet

A smooth, nearly flat surface at one end of bone, which articulates with another bone.

Condyle

A large rounded projection which articulates with another bone (found at the knee joint).

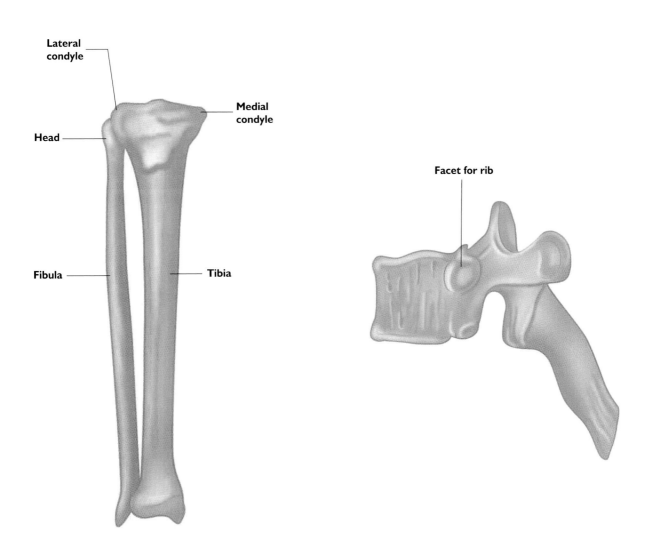

Figure 48: Projections on bones that help to form joints.

3. Depressions and Openings that Allow Blood Vessels and Nerves to Pass Through

Sinus
A cavity within a bone that is filled with air and lined with a membrane (most notably in the skull).

Fossa
A shallow, basin-like depression in a bone, often serving as an articular surface.

Foramen (pl. foramina)
A round or oval opening through a bone (most notably on the sacrum).

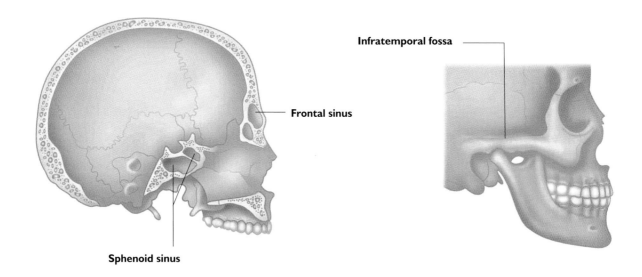

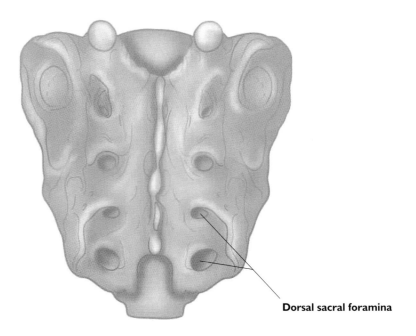

Figure 49: Depressions and openings that allow blood vessels and nerves to pass through.

The Axial Skeleton

4

The Skull:
Comprising the
Cranium and the
Facial Bones

The Vertebral
Column (Spine)

The Bony Thorax

The Skull: Comprising the Cranium and the Facial Bones

NOTE: Some soft tissues (e.g. cartilages, aponeuroses, ligaments and tendons) are included where appropriate, for ease of reference.

The Cranium

Eight large flat bones: comprising two pairs, plus four single bones. These form a box-like container that houses the brain. These bones are:

Frontal: which forms the forehead, the bony projections under the eyebrows and the superior part of each eye orbit.

Parietal: a pair of bones that form most of the superior and lateral walls of the cranium. They meet in the midline at the *sagittal suture*, and meet with the frontal bone at the *coronal suture*.

Temporal: a pair of bones which lie inferior to the parietal bones; there are three important markings on the temporal bone: (a) the *styloid process* which is just in front of the mastoid process; a sharp needle-like projection to which many of the neck muscles attach; (b) the *zygomatic process*, a thin bridge of bone that joins with the zygomatic bone just above the mandible; (c) the *mastoid process*, a rough projection posterior and inferior to the styloid process (just behind the lobe of the ear).

Occipital: the most posterior bone of the cranium. It forms the floor and back wall of the skull; and joins the parietal bones anteriorly at the *lambdoidal suture*. In the base of the occipital bone is a large opening, the *foramen magnum* through which the spinal cord passes to connect with the brain. To each side of the foramen magnum are the *occipital condyles* that rest on the first vertebra of the spinal column (the atlas).

Sphenoid: a butterfly-shaped bone that spans the width of the skull and forms part of the floor of the cranial cavity. Parts of the sphenoid can be seen forming part of the eye orbits, and the lateral part of the skull.

Ethmoid (*see* figure 53)**:** a single bone in front of the sphenoid bone and below the frontal bone. Forms part of nasal septum and superior and medial conchae.

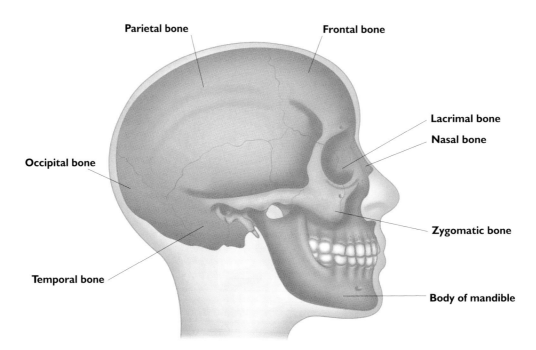

Figure 50: Skull (lateral view).

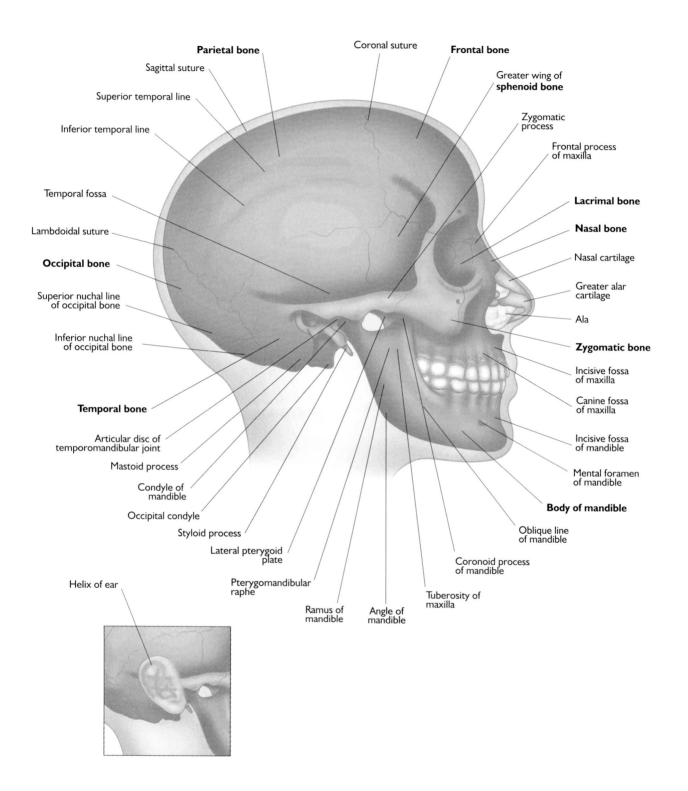

Parietal bone

Coronal suture

Frontal bone

Sagittal suture

Greater wing of
sphenoid bone

Superior temporal line

Zygomatic
process

Inferior temporal line

Frontal process
of maxilla

Temporal fossa

Lacrimal bone

Lambdoidal suture

Nasal bone

Occipital bone

Nasal cartilage

Superior nuchal line
of occipital bone

Greater alar
cartilage

Inferior nuchal line
of occipital bone

Ala

Zygomatic bone

Temporal bone

Incisive fossa
of maxilla

Articular disc of
temporomandibular joint

Canine fossa
of maxilla

Mastoid process

Incisive fossa
of mandible

Condyle of
mandible

Mental foramen
of mandible

Occipital condyle

Body of mandible

Styloid process

Oblique line
of mandible

Lateral pterygoid
plate

Coronoid process
of mandible

Pterygomandibular
raphe

Tuberosity of
maxilla

Helix of ear

Ramus of
mandible

Angle of
mandible

Figure 51: Skull (lateral view).

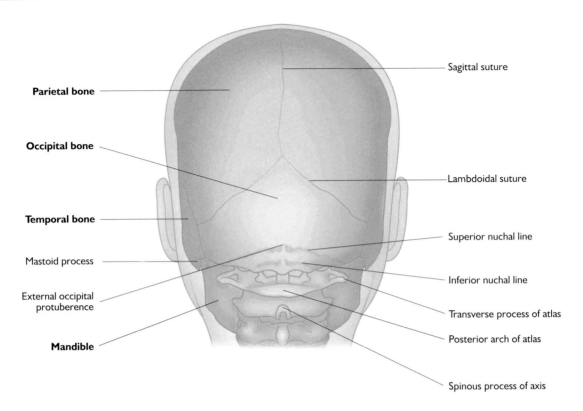

Parietal bone

Occipital bone

Temporal bone

Mastoid process

External occipital protuberence

Mandible

Sagittal suture

Lambdoidal suture

Superior nuchal line

Inferior nuchal line

Transverse process of atlas

Posterior arch of atlas

Spinous process of axis

Figure 52: Skull (posterior view).

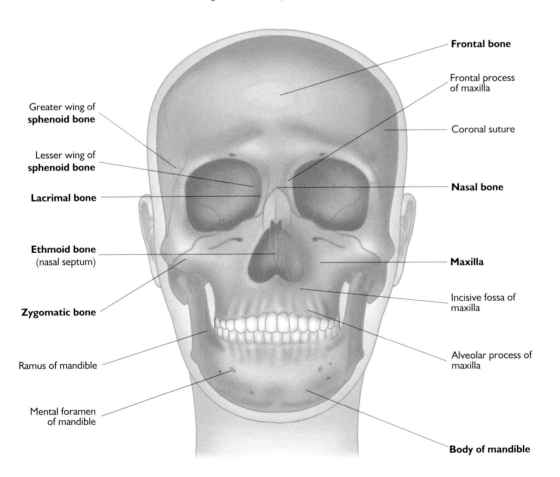

Greater wing of sphenoid bone

Lesser wing of sphenoid bone

Lacrimal bone

Ethmoid bone (nasal septum)

Zygomatic bone

Ramus of mandible

Mental foramen of mandible

Frontal bone

Frontal process of maxilla

Coronal suture

Nasal bone

Maxilla

Incisive fossa of maxilla

Alveolar process of maxilla

Body of mandible

Figure 53: Skull (anterior view).

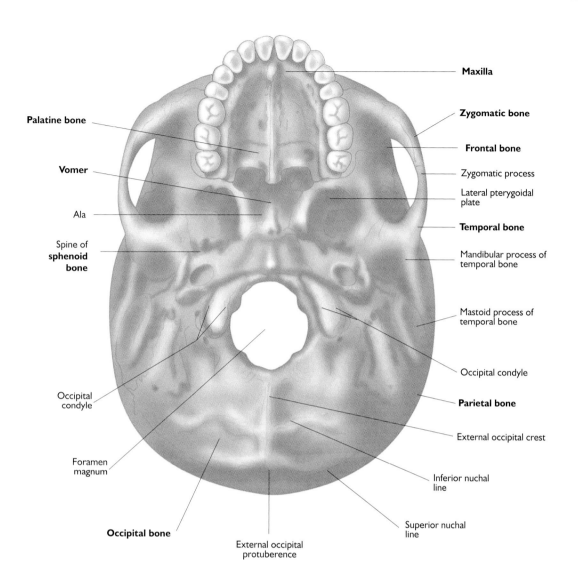

Maxilla

Zygomatic bone

Frontal bone

Zygomatic process

Lateral pterygoidal plate

Temporal bone

Mandibular process of temporal bone

Mastoid process of temporal bone

Occipital condyle

Parietal bone

External occipital crest

Inferior nuchal line

Superior nuchal line

Palatine bone

Vomer

Ala

Spine of sphenoid bone

Occipital condyle

Foramen magnum

Occipital bone

External occipital protuberence

Figure 54: Skull (basal view).

The Facial Bones

Fourteen bones compose the face, twelve of which are pairs. The main bones of the face are:

Nasal: a pair of small rectangular bones that form the bridge of the nose (the lower part of the nose is made up of cartilage).

Zygomatic: a pair of bones commonly known as the cheekbones. They also form a large portion of the lateral walls of the eye orbits.

Maxillae: the two maxillary bones fuse to form the upper jaw. The upper teeth are imbedded in the maxillae.

Mandible: the lower jawbone is the strongest bone in the face; it joins the temporal bones on each side of the face, forming the only freely movable joints in the skull. The horizontal part of the mandible, or the *body*, forms the chin. Two upright bars of bone, or *rami*, extend from the body to connect the mandible with the temporal bone. The lower teeth are imbedded in the mandible.

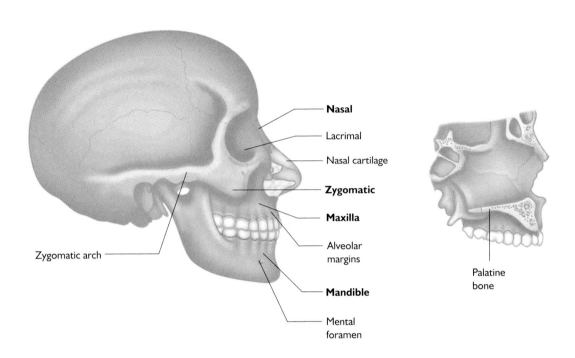

Nasal

Lacrimal

Nasal cartilage

Zygomatic

Maxilla

Alveolar margins

Mandible

Zygomatic arch

Mental foramen

Palatine bone

Figure 55: Facial bones (lateral view).

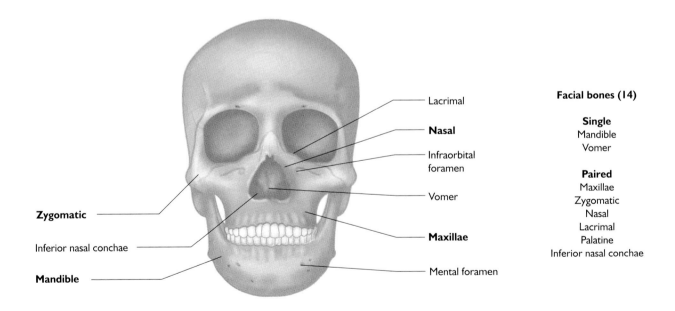

Lacrimal

Nasal

Infraorbital foramen

Vomer

Maxillae

Zygomatic

Inferior nasal conchae

Mandible

Mental foramen

Facial bones (14)

Single
Mandible
Vomer

Paired
Maxillae
Zygomatic
Nasal
Lacrimal
Palatine
Inferior nasal conchae

Figure 56: Facial bones (anterior view).

The Vertebral Column (Spine)

The vertebral column consists of 33 vertebrae in total:

- 7 cervical vertebrae.
- 12 thoracic vertebrae – which also form joints with the 12 ribs.
- 5 lumbar vertebrae – the largest, weight-bearing vertebrae.
- Sacrum (5 fused) – note that the holes, or foramina, in the sacrum correspond to the original gaps between the vertebrae.
- Coccyx (3–4 fused).

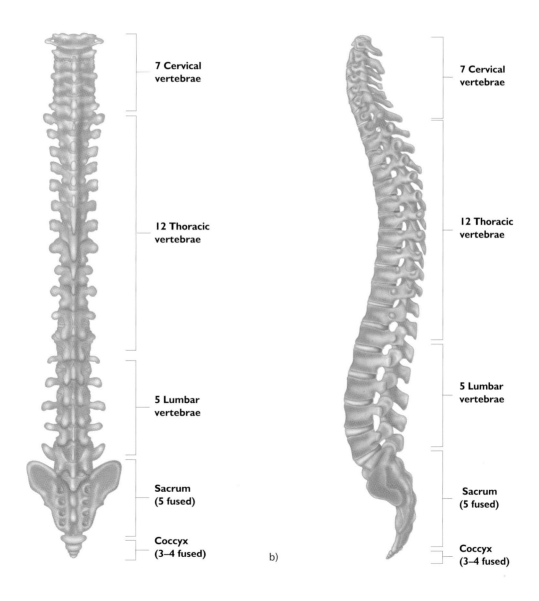

Figure 57: The vertebral column; a) posterior view, b) lateral view.

A Typical Vertebra

A typical vertebra has the following parts:

Body: the disk-like, weight-bearing part of the vertebra. It faces anteriorly in the vertebral column.

Vertebral arch: the arch formed from the joining of the processes to the body.

Vertebral foramen: the canal through which the spinal cord passes.

Transverse process: two lateral projections.

Spinous process: a single projection that rises from the posterior part of the vertebral arch. On the cervical vertebrae, the spinous processes are short and divide into two points (it looks a little like a whale's tail). On the thoracic vertebrae, the spinous processes are single, slender points that angle sharply downward. On the lumbar vertebrae, the spinous processes are thick and wedge shaped.

Superior and inferior articular processes: paired projections lateral to the foramen. They allow one vertebra to form a joint with the next vertebra.

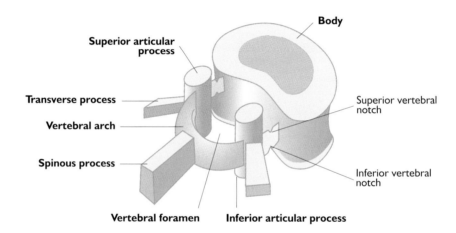

Figure 58: A typical vertebra (schematic).

The illustrations on the following pages give a selection of key vertebrae shown from various angles, to depict their variation in shape and features.

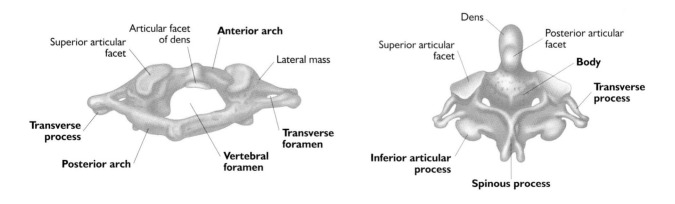

Figure 59: Atlas (C1) postero-superior view.

Figure 60: Axis (C2) postero-superior view.

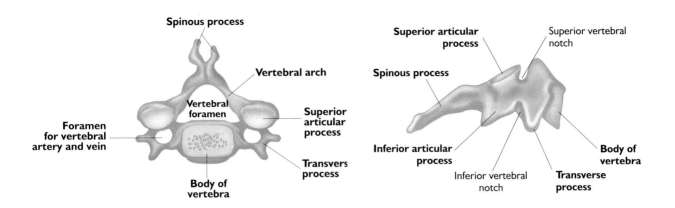

Figure 61a: Cervical vertebra (C5) superior view.

Figure 61b: Cervical vertebra (C5) lateral view.

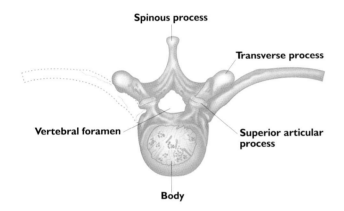

Figure 62a: Thoracic vertebra (T6) superior view.

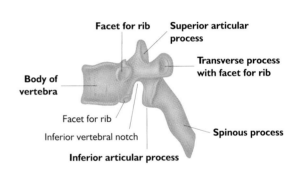

Figure 62b: Thoracic vertebra (T6) lateral view.

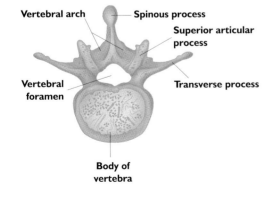

Figure 63a: Lumbar vertebra (L3) superior view.

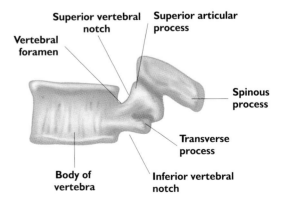

Figure 63b: Lumbar vertebra (L3) lateral view.

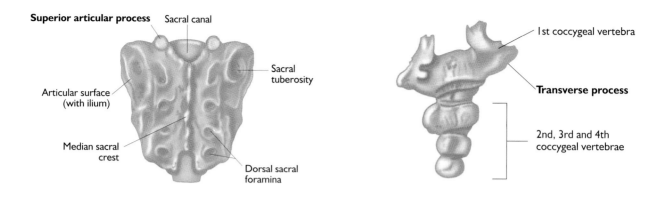

Figure 64: The sacrum: posterior view.

Figure 65: The coccyx: posterior view.

The Bony Thorax

Sternum

The sternum is commonly known as the breastbone. It is actually the fusion of three bones: the manubrium, the body (also known as the *gladiolus*), and the xiphoid process.

NOTE: The sternum is attached to the first seven pairs of ribs by the costal cartilage; Manubrium means '*handle*', as in the handle of a sword; Xiphoid means '*sword shaped*'.

The Ribs

The ribs consist of 12 pairs in total (comprising true, false and floating ribs).

- True ribs: the first seven pairs attach by costal cartilage directly to the sternum.
- False ribs: the next three pairs attach to costal cartilage but not directly to the sternum.
- Floating ribs: the last two pairs of ribs lack attachment either to costal cartilage or to the sternum.

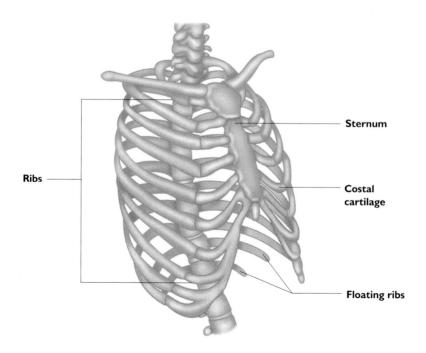

Figure 66: The ribs and sternum.

The Appendicular Skeleton

5

NOTE: Some soft tissues (e.g. cartilages, aponeuroses, ligaments and tendons) are included where appropriate, for ease of reference.

The Pectoral Girdle

Clavicle
The clavicle is commonly known as the collarbone; a slender, doubly curved bone that attaches to the manubrium of the sternum medially (the *sternoclavicular joint*) and to the acromion of the scapula laterally (the *acromioclavicular joint*).

Scapula
Commonly known as the shoulder blade; the scapula is a large triangular flat bone lying posterior to the dorsal thorax between the second and seventh ribs. Each scapula articulates with the clavicle and the humerus. The scapula has four important bone markings:

1. The spine – a sharp, prominent ridge on the posterior surface of the scapula which can be easily felt through the skin.
2. The acromion – an enlarged anterior projection at the lateral end of the spine of scapula, that can be felt as the 'point of the shoulder'.
3. The corocoid process – projecting forward from the upper border of the scapula.
4. The glenoid fossa – a shallow depression at the lateral angle of the scapula that articulates with the head of the humerus.

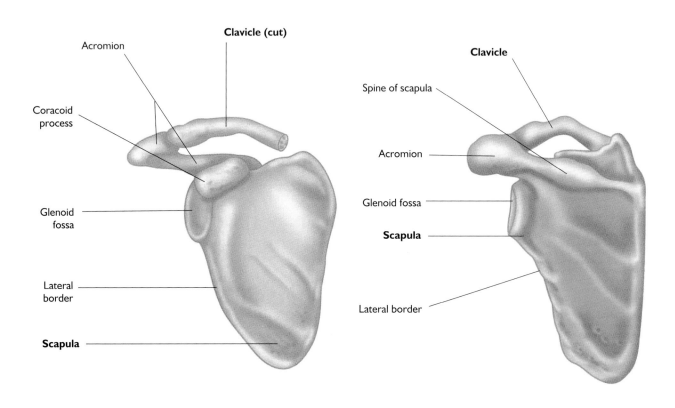

Figure 67a: The clavicle and scapula (anterior view).   Figure 67b: The scapula (posterior view).

The Upper Limb

Humerus

The humerus (arm bone) is the longest and largest bone of the upper limb. It articulates proximally with the scapula (at the glenoid fossa). At the distal end are the *trochlea* (which looks like a spool) and the *capitulum* (or head), which form part of the elbow joint with the ulna and the radius. On either side of the trochlea are the medial and lateral epicondyles of the humerus, easily felt superficially.

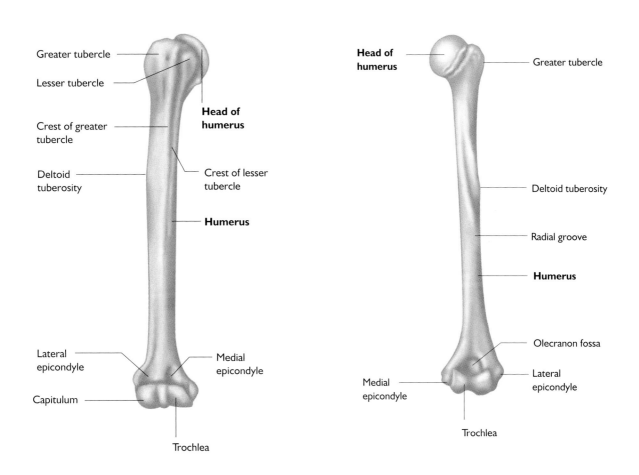

Figure 68: The right humerus (anterior view). Figure 69: The right humerus (posterior view).

Radius

The radius is one of the two bones in the forearm, on the lateral, or thumb side of the forearm. Proximally the head of the radius forms a joint with the capitulum of the humerus. The radius crosses the ulna during pronation.

Ulna

The ulna is the medial bone in the forearm, on the little finger side. At the proximal end of the ulna are two processes: the *coronoid* and the *olecranon*, which fit over the two medial rounded spools of the trochlea of the humerus. The olecranon is the pointed bump felt when the elbow is bent, and is also known as 'the funny bone', because when the nerve that runs over the olecranon is hit, it can be painful. The *styloid processes* of the radius and ulna can be felt as sharp projections on either side of the wrist.

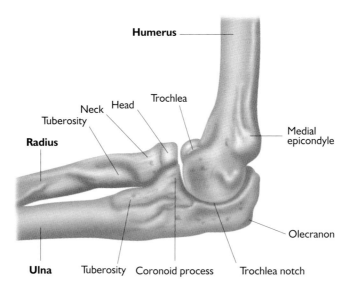

Figure 70: Right elbow: medial view in 90° flexion.

8 Carpals

The eight small carpal bones make up the wrist. They are bound together by ligaments and are arranged in two transverse rows, four bones to a row. The first row comprises the scaphoid, lunate, triquetrum, pisiform. The second row comprises the trapezium, trapezoid, capitate, hamate. A mnemonic for memorizing the carpals from lateral to medial, beginning with the proximal row, is: "some lovers try positions that they can't handle".

5 Metacarpals

The metacarpals are five bones running between the wrist and the knuckles (which are the heads of the metacarpals).

14 Phalanges

Each finger has three phalanges, whereas the thumb only has two.

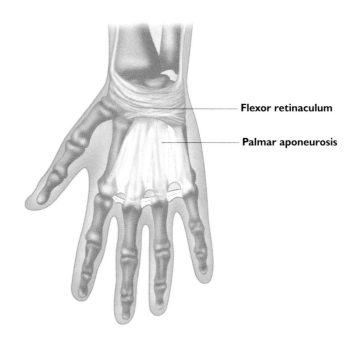

Figure 71: The hand (anterior view).

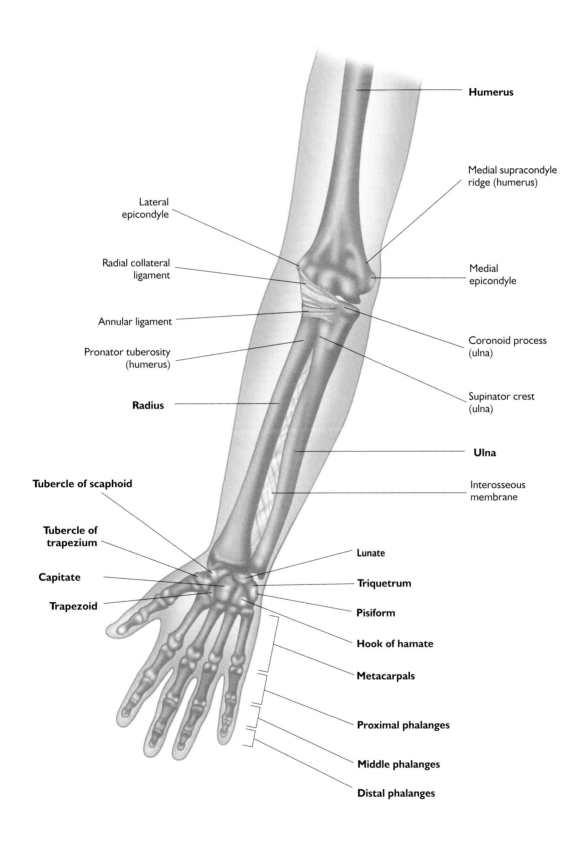

Humerus

Medial supracondyle
ridge (humerus)

Lateral
epicondyle

Radial collateral
ligament

Medial
epicondyle

Annular ligament

Pronator tuberosity
(humerus)

Coronoid process
(ulna)

Supinator crest
(ulna)

Radius

Ulna

Interosseous
membrane

Tubercle of scaphoid

Tubercle of
trapezium

Lunate

Capitate

Triquetrum

Trapezoid

Pisiform

Hook of hamate

Metacarpals

Proximal phalanges

Middle phalanges

Distal phalanges

Figure 72: The bones of the right forearm and hand (anterior view).

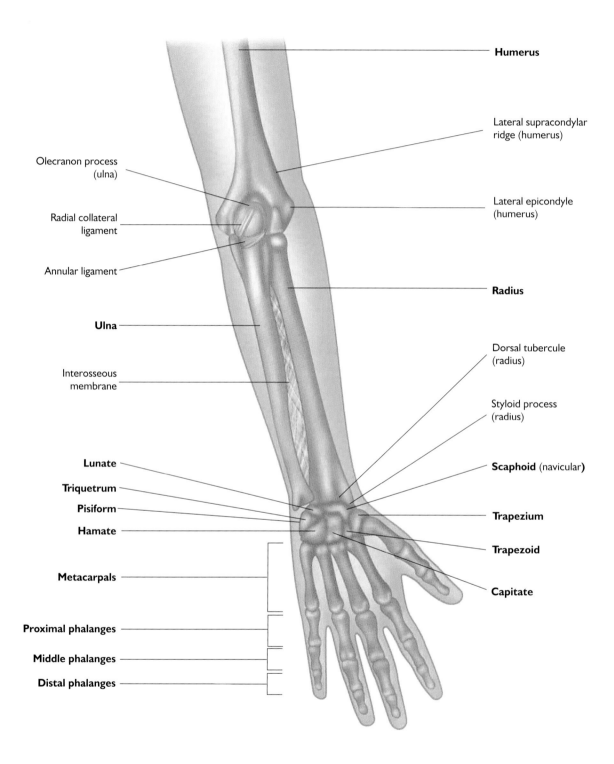

Olecranon process
(ulna)

Radial collateral
ligament

Annular ligament

Ulna

Interosseous
membrane

Lunate

Triquetrum

Pisiform

Hamate

Metacarpals

Proximal phalanges

Middle phalanges

Distal phalanges

Humerus

Lateral supracondylar
ridge (humerus)

Lateral epicondyle
(humerus)

Radius

Dorsal tubercule
(radius)

Styloid process
(radius)

Scaphoid (navicular**)**

Trapezium

Trapezoid

Capitate

Figure 73: The bones of the right forearm and hand (posterior view).

The Pelvic Girdle (Os Innominatum)

The pelvic girdle (or hip girdle) consists of two pelvic, or coxal bones. It provides a strong and stable support for the lower extremities on which the weight of the body is carried. The pelvic bones unite with one another in the front (anteriorly) at the *pubic symphysis* (a fibrocartilage disc). With the sacrum and coccyx, the two pelvic bones form a basin-like structure called the pelvis. At birth, each pelvic or coxal bone consists of three separate bones; the ilium, the ischium, and the pubis. These separate bones eventually fuse into one pelvic bone, and the area where they join is a deep hemispherical socket called the *acetabulum* (this socket articulates with the head of the femur). Although the pelvic bone is one bone, it is still commonly discussed as if it consisted of three portions.

Ilium

The ilium is a large, flaring bone that forms the largest and most superior portion of the pelvic bone. *Iliac crests* are felt when you rest your hands on your hips. Each crest terminates in the front as the *Anterior Superior Iliac Spine* or ASIS; and at the back as the *Posterior Superior Iliac Spine* or PSIS (the PSIS is difficult to palpate, but its position is revealed by a skin dimple in the sacral region, level approximately to the second sacral foramen).

Ischium

The ischium is the inferior, posterior part of the pelvic bone, roughly arch-shaped. At the bottom of the ischium are the roughened and thickened *ischial tuberosities* (sometimes called the 'sit-bones', because when we sit, our weight is borne entirely by the ischial tuberosities).

Pubis

The pubis is the anterior and inferior part of the pelvic bone.

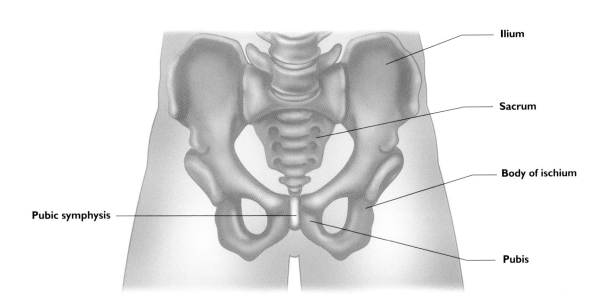

Fig 74a: Bones of the pelvic girdle (anterior view).

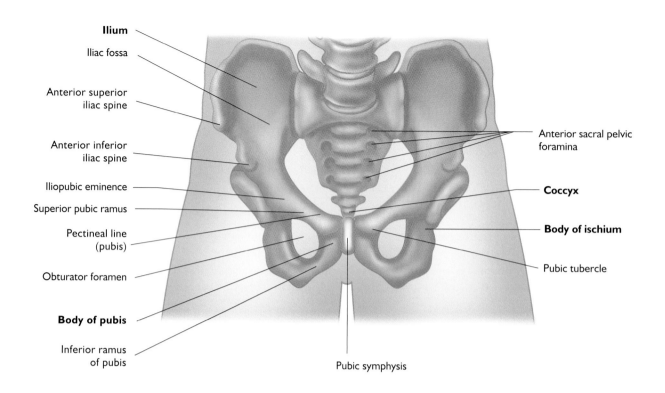

Ilium
Iliac fossa
Anterior superior iliac spine
Anterior inferior iliac spine
Iliopubic eminence
Superior pubic ramus
Pectineal line (pubis)
Obturator foramen
Body of pubis
Inferior ramus of pubis

Anterior sacral pelvic foramina
Coccyx
Body of ischium
Pubic tubercle
Pubic symphysis

Figure 74b: Bones of the pelvic girdle (anterior view).

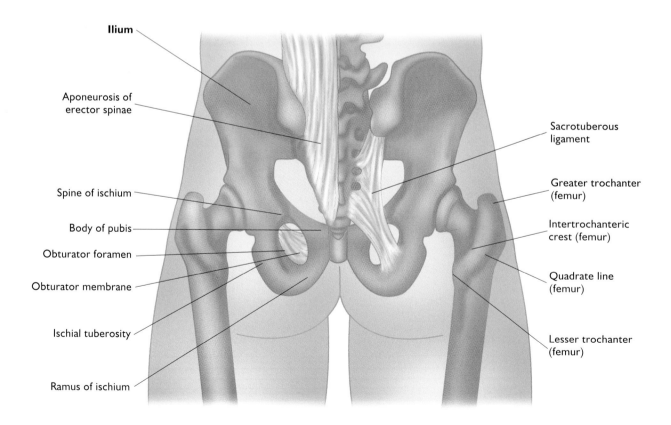

Ilium
Aponeurosis of erector spinae
Spine of ischium
Body of pubis
Obturator foramen
Obturator membrane
Ischial tuberosity
Ramus of ischium

Sacrotuberous ligament
Greater trochanter (femur)
Intertrochanteric crest (femur)
Quadrate line (femur)
Lesser trochanter (femur)

Figure 75: Bones of the pelvic girdle (posterior view).

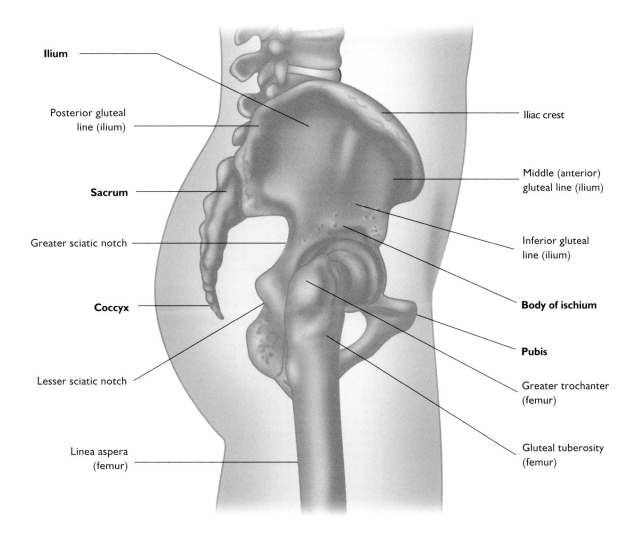

Ilium

Posterior gluteal
line (ilium)

Sacrum

Greater sciatic notch

Coccyx

Lesser sciatic notch

Linea aspera
(femur)

Iliac crest

Middle (anterior)
gluteal line (ilium)

Inferior gluteal
line (ilium)

Body of ischium

Pubis

Greater trochanter
(femur)

Gluteal tuberosity
(femur)

Figure 76: Bones of the pelvic girdle (lateral view).

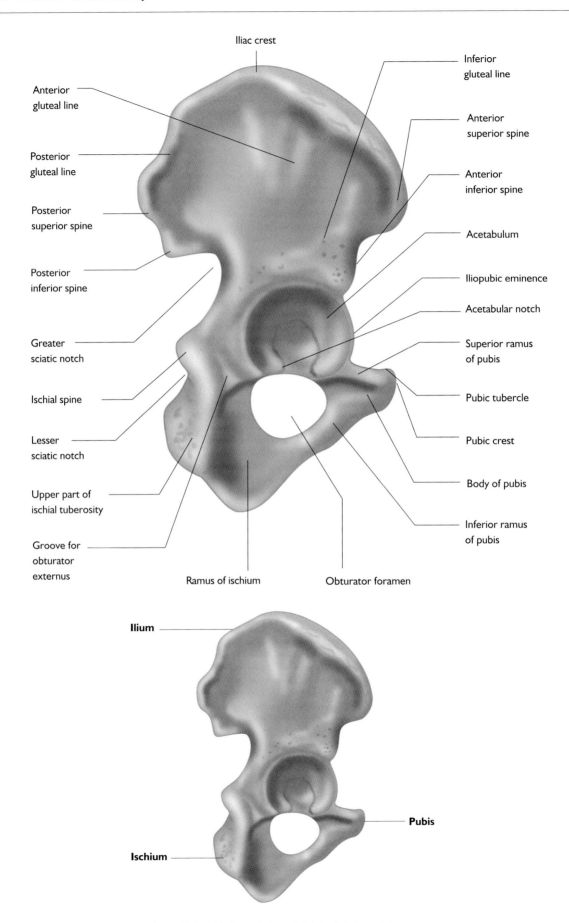

Iliac crest

Inferior gluteal line

Anterior gluteal line

Anterior superior spine

Posterior gluteal line

Anterior inferior spine

Posterior superior spine

Acetabulum

Posterior inferior spine

Iliopubic eminence

Acetabular notch

Greater sciatic notch

Superior ramus of pubis

Ischial spine

Pubic tubercle

Lesser sciatic notch

Pubic crest

Upper part of ischial tuberosity

Body of pubis

Groove for obturator externus

Inferior ramus of pubis

Ramus of ischium

Obturator foramen

Ilium

Pubis

Ischium

Figure 77 (a & b): Lateral view of right pelvic (coxal) bone.

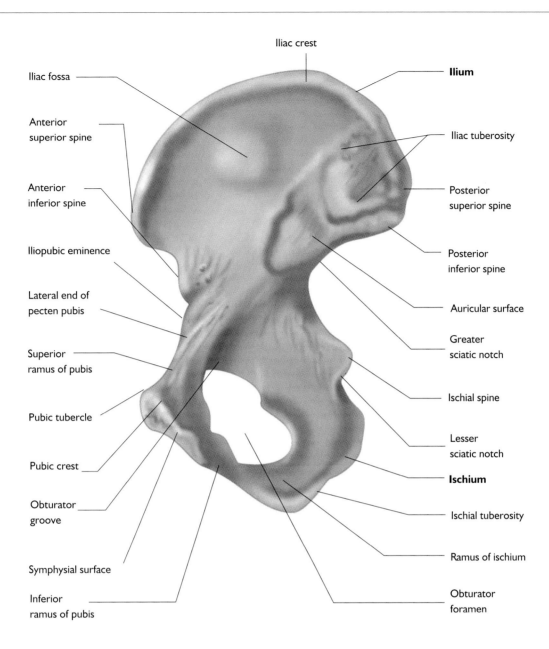

Iliac crest

Iliac fossa

Anterior
superior spine

Anterior
inferior spine

Iliopubic eminence

Lateral end of
pecten pubis

Superior
ramus of pubis

Pubic tubercle

Pubic crest

Obturator
groove

Symphysial surface

Inferior
ramus of pubis

Ilium

Iliac tuberosity

Posterior
superior spine

Posterior
inferior spine

Auricular surface

Greater
sciatic notch

Ischial spine

Lesser
sciatic notch

Ischium

Ischial tuberosity

Ramus of ischium

Obturator
foramen

Figure 78: Bones of the pelvic girdle (posterior view).

The Lower Limb

Femur

The femur is the only bone in the thigh. It is the heaviest, longest and strongest bone in the body. Its proximal end has a ball-like head that articulates with the pelvic bone at the acetabulum. Distally on the femur are the lateral and medial condyles, which articulate with the tibia.

– the *greater trochanter* is a projection just distal to the head and neck of the femur and can sometimes be felt in the buttock.

Tibia

The tibia (shin bone) is the larger and more medial of the bones in the lower leg. At the proximal end, the *medial and lateral condyles* articulate with the distal end of the femur to form the knee joint.

– the *tibial tuberosity* is a roughened area on the anterior surface of the tibia.
– the *medial malleolus* can be felt as the inner bulge of the ankle.

Fibula

The fibula lies lateral and parallel to the tibia and is thin and sticklike. The fibula is not a weight bearing bone. It also plays no part in the knee joint.

– the *lateral malleolus* on the fibula can be felt as the outer bulge of the ankle.

Patella

Known as the 'knee cap', the patella is a small triangular sesamoid bone within the tendon of the quadriceps femoris muscle. It forms the front of the knee joint.

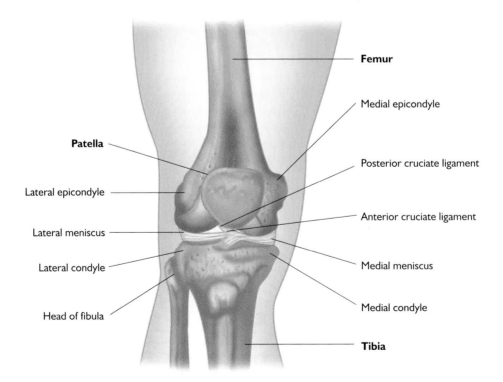

Figure 79: Lower femur and upper tibia of the right leg (anterior view).

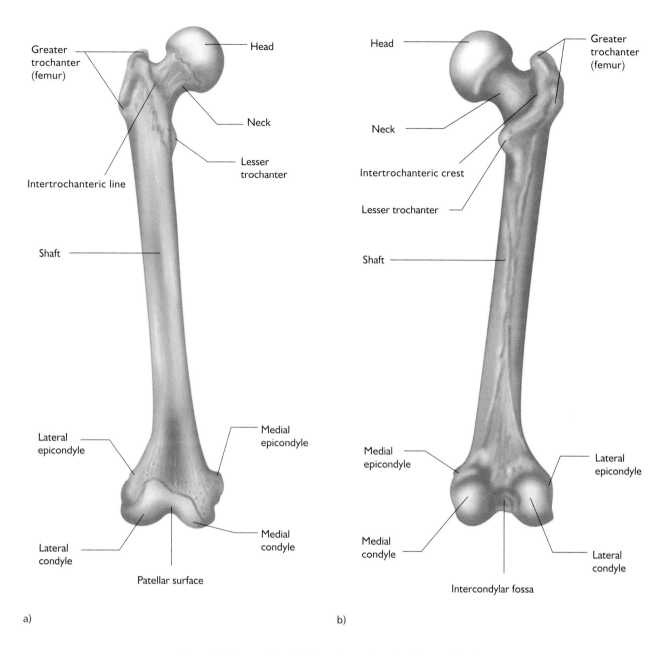

Greater trochanter (femur)

Head

Intertrochanteric line

Neck

Lesser trochanter

Shaft

Lateral epicondyle

Medial epicondyle

Lateral condyle

Medial condyle

Patellar surface

a)

Head

Greater trochanter (femur)

Neck

Intertrochanteric crest

Lesser trochanter

Shaft

Medial epicondyle

Lateral epicondyle

Medial condyle

Lateral condyle

Intercondylar fossa

b)

Figure 80: Femur of the right leg; a) anterior view, b) posterior view.

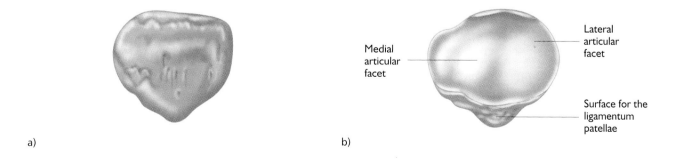

Medial articular facet

Lateral articular facet

Surface for the ligamentum patellae

a)

b)

Figure 81: Patella of the right leg; a) anterior view, b) posterior view.

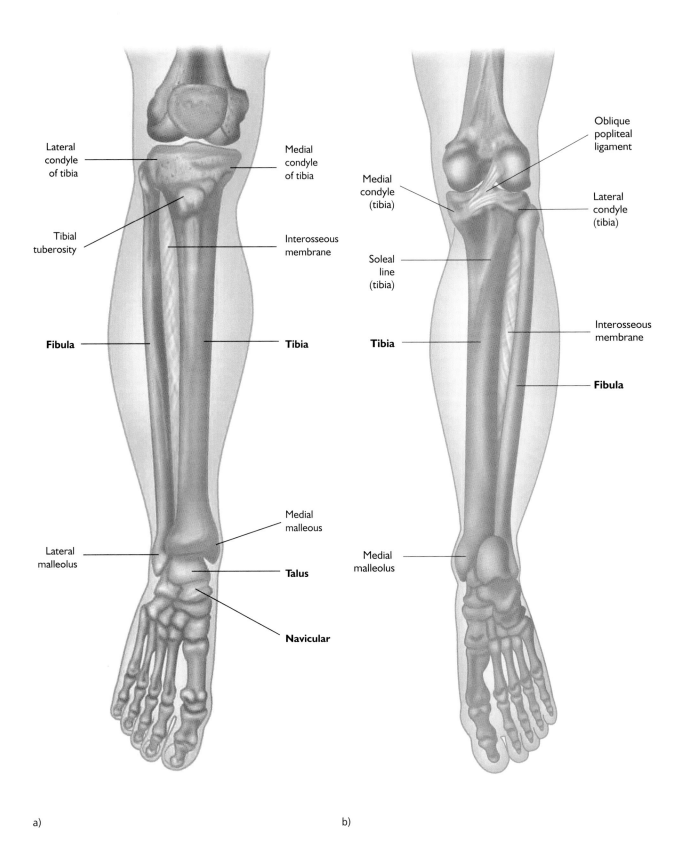

Lateral
condyle
of tibia

Tibial
tuberosity

Fibula

Medial
condyle
of tibia

Interosseous
membrane

Tibia

Medial
malleous

Lateral
malleolus

Talus

Navicular

Oblique
popliteal
ligament

Medial
condyle
(tibia)

Lateral
condyle
(tibia)

Soleal
line
(tibia)

Tibia

Interosseous
membrane

Fibula

Medial
malleolus

a)   b)

Figure 82: Tibia and fibula of the right leg, a) anterior view, b) posterior view.

7 Tarsals

The tarsals are the seven bones of the ankle. The two largest tarsals mostly carry the body weight: the *calcaneus*, or the heelbone, and the *talus*, which lies between the tibia and the calcaneus. The navicular, medial cuneiform, intermediate cuneiform, lateral cuneiform and cuboid constitute the other five tarsals.

5 Metatarsals

The metatarsals form the instep or sole of the foot.

14 Phalanges

Each toe has three phalanges, except the big toe, which has only two.

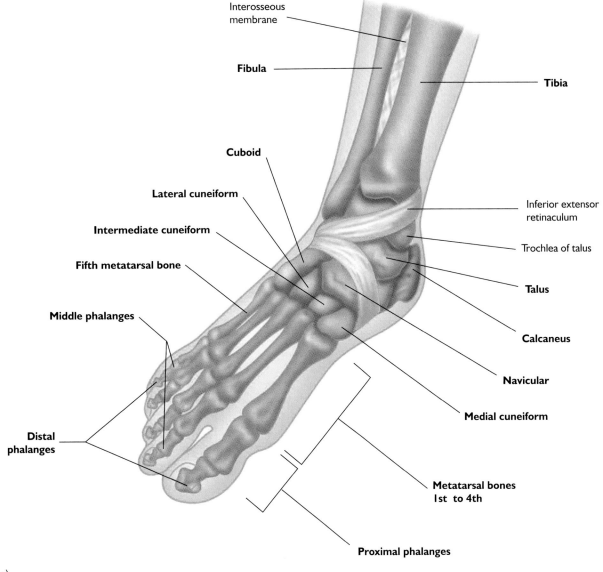

a)

Figure 83: Bones of the right foot; a) anteromedial view, b) lateral view, c) plantar view.

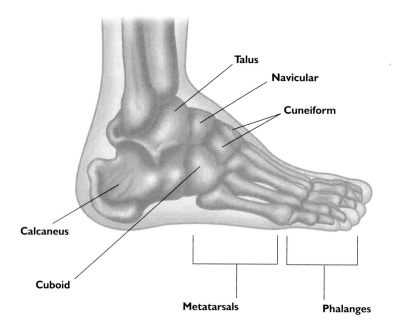

Talus

Navicular

Cuneiform

Calcaneus

Cuboid

Metatarsals

Phalanges

b)

Medial
malleolus
of tibia

Tuberosity
of calcaneus

Medial border
of calcaneus

Talus

Sustentaculum
tali of calcaneus

Tuberosity
of navicular

Lateral border
of calcaneus

**Intermediate
cuneiform**

Cuboid

**Medial
cuneiform**

**Lateral
cuneiform**

**Metatarsal
bones**

**Middle
phalanges**

**Proximal
phalanges**

**Distal
phalanges**

c)

Figure 83 (cont.): Bones of the right foot; a) anteromedial view, b) lateral view, c) plantar view.

General Skeletal Interrelationships

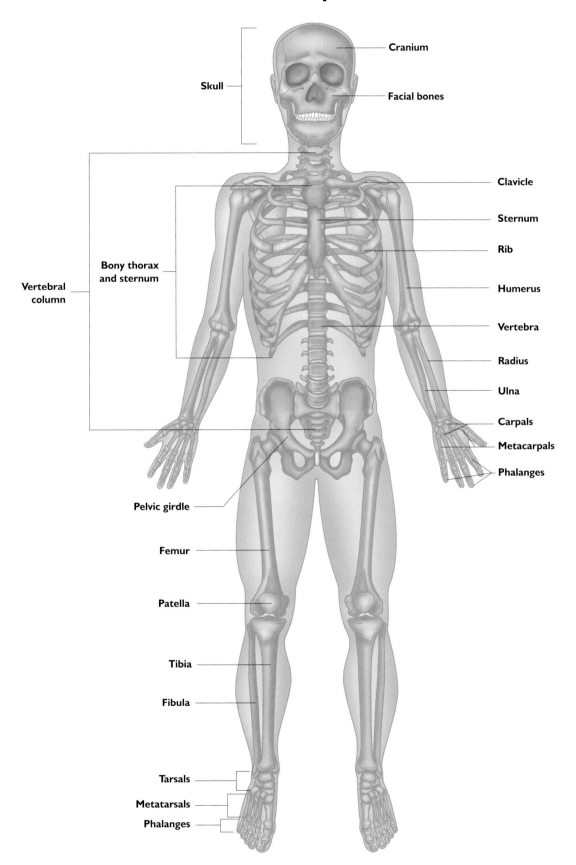

Figure 84: Skeleton (anterior view).

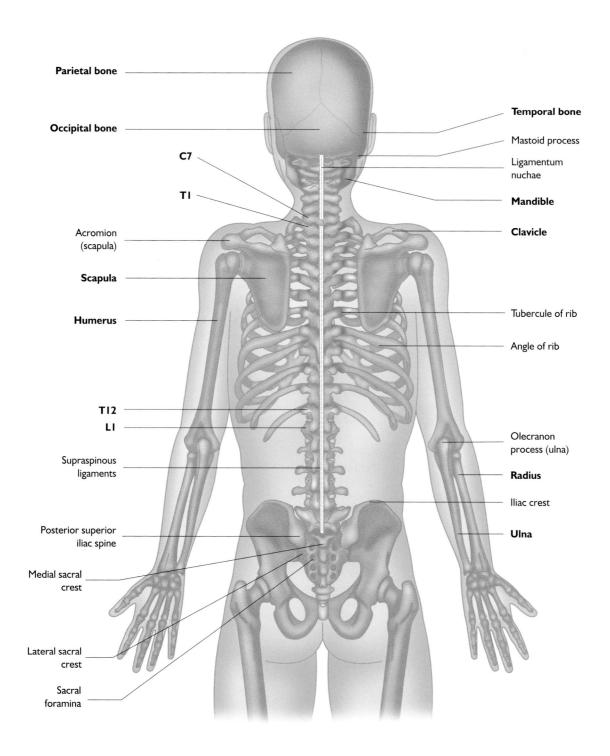

Parietal bone

Occipital bone

C7

T1

Acromion
(scapula)

Scapula

Humerus

T12

L1

Supraspinous
ligaments

Posterior superior
iliac spine

Medial sacral
crest

Lateral sacral
crest

Sacral
foramina

Temporal bone

Mastoid process

Ligamentum
nuchae

Mandible

Clavicle

Tubercule of rib

Angle of rib

Olecranon
process (ulna)

Radius

Iliac crest

Ulna

Figure 85: Skeleton (posterior view).

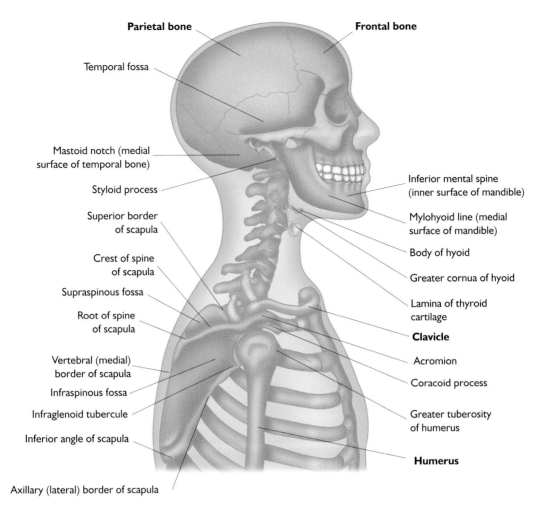

Parietal bone

Frontal bone

Temporal fossa

Mastoid notch (medial surface of temporal bone)

Styloid process

Superior border of scapula

Crest of spine of scapula

Supraspinous fossa

Root of spine of scapula

Vertebral (medial) border of scapula

Infraspinous fossa

Infraglenoid tubercule

Inferior angle of scapula

Axillary (lateral) border of scapula

Inferior mental spine (inner surface of mandible)

Mylohyoid line (medial surface of mandible)

Body of hyoid

Greater cornua of hyoid

Lamina of thyroid cartilage

Clavicle

Acromion

Coracoid process

Greater tuberosity of humerus

Humerus

Figure 86: Skull to humerus (lateral view).

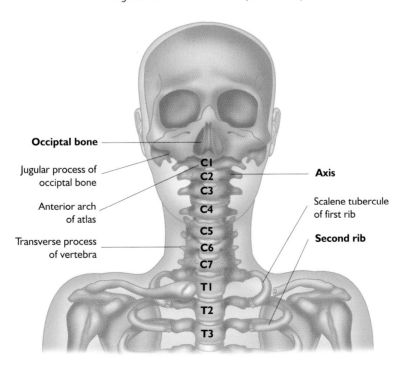

Occiptal bone

Jugular process of occiptal bone

Anterior arch of atlas

Transverse process of vertebra

C1
C2
C3
C4
C5
C6
C7
T1
T2
T3

Axis

Scalene tubercule of first rib

Second rib

Figure 87: Skull to sternum (anterior view, the mandible and maxilla are removed).

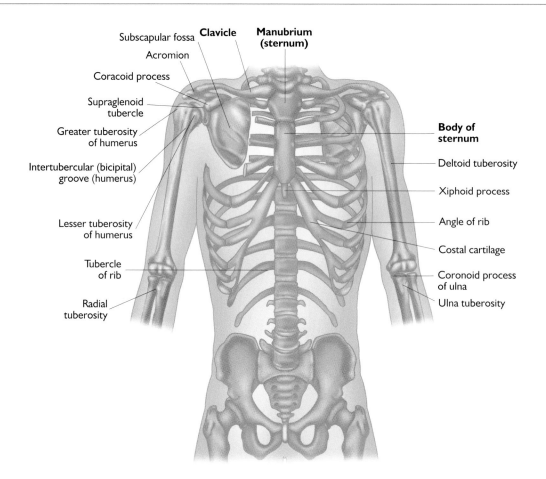

Subscapular fossa

Clavicle

Manubrium (sternum)

Acromion

Coracoid process

Supraglenoid tubercle

Greater tuberosity of humerus

Body of sternum

Intertubercular (bicipital) groove (humerus)

Deltoid tuberosity

Xiphoid process

Lesser tuberosity of humerus

Angle of rib

Costal cartilage

Tubercle of rib

Coronoid process of ulna

Radial tuberosity

Ulna tuberosity

Figure 88: Ribcage, pectoral girdle, upper arm (anterior view, the upper right anterior ribcage is removed).

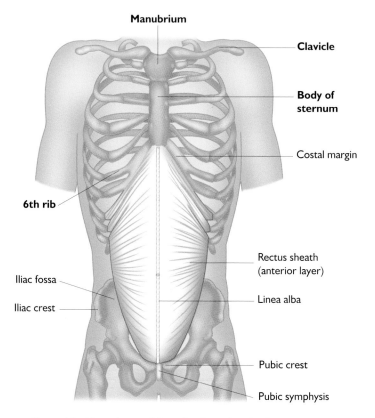

Manubrium

Clavicle

Body of sternum

Costal margin

6th rib

Rectus sheath (anterior layer)

Iliac fossa

Linea alba

Iliac crest

Pubic crest

Pubic symphysis

Figure 89a: Thoracic to pelvic region (anterior view).

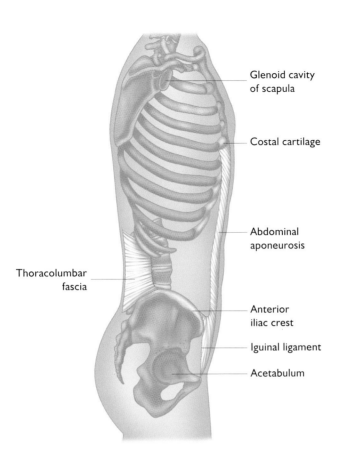

Glenoid cavity
of scapula

Costal cartilage

Abdominal
aponeurosis

Thoracolumbar
fascia

Anterior
iliac crest

Iguinal ligament

Acetabulum

Figure 89b: Thoracic to pelvic region (lateral view).

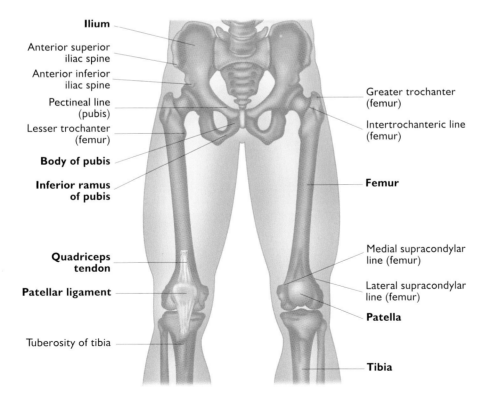

Ilium

Anterior superior
iliac spine

Anterior inferior
iliac spine

Pectineal line
(pubis)

Lesser trochanter
(femur)

Body of pubis

**Inferior ramus
of pubis**

**Quadriceps
tendon**

Patellar ligament

Tuberosity of tibia

Greater trochanter
(femur)

Intertrochanteric line
(femur)

Femur

Medial supracondylar
line (femur)

Lateral supracondylar
line (femur)

Patella

Tibia

Figure 90a: Pelvic girdle to leg (anterior view).

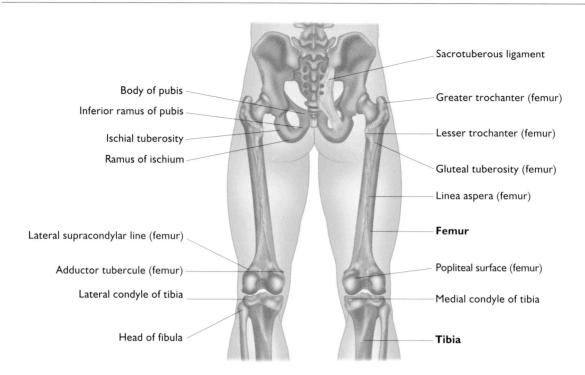

Sacrotuberous ligament

Body of pubis

Inferior ramus of pubis

Greater trochanter (femur)

Ischial tuberosity

Lesser trochanter (femur)

Ramus of ischium

Gluteal tuberosity (femur)

Linea aspera (femur)

Femur

Lateral supracondylar line (femur)

Adductor tubercule (femur)

Popliteal surface (femur)

Lateral condyle of tibia

Medial condyle of tibia

Head of fibula

Tibia

Figure 90b: Pelvic girdle to leg (posterior view).

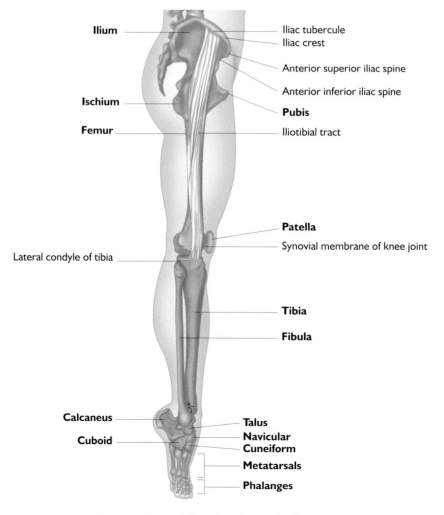

Ilium

Iliac tubercule
Iliac crest

Anterior superior iliac spine

Anterior inferior iliac spine

Ischium

Pubis

Femur

Iliotibial tract

Patella

Synovial membrane of knee joint

Lateral condyle of tibia

Tibia

Fibula

Calcaneus

Talus

Cuboid

Navicular

Cuneiform

Metatarsals

Phalanges

Figure 91: Pelvic girdle to foot (lateral view).

Bony Landmarks Seen or Felt Near the Body Surface

The following bony landmarks can be seen or felt near the surface of the body. Identify them on yourself or a partner using figure 92 (a–c) for reference.

Frontal bone
Temporal bone
Occipital bone
Manubriosternum and
    manubriosternal joint
    (level with the 2nd rib)
2nd rib
Sternoclavicular joint
Acromioclavicular joint
Spine of the scapula
Medial border of scapula
Inferior angle of scapula
Medial and lateral epicondyle
    of humerus
Olecranon
Head of the radius
Ulnar styloid
Pisiform bone
Anatomical snuff box
Iliac crest
Anterior superior iliac spine (ASIS)
Posterior superior iliac spine (PSIS)
Ischial tuberosities
Greater trochanter
Head of the fibula
Tibial tuberosity
Medial and lateral malleolus
Calcaneus
Spinous process of the vertebrae

Hints

C2: the first cervical vertebra
    to be felt below the occiput.
C7: at the base of the neck,
    the vertebra that stands out
    most prominently.
T3–4: level with the spine
    of the scapula.
T7: level with the inferior
    angle of the scapula.
L4: level with the iliac crest.
S2: level with the PSIS
    (or visible as the dimple
    at the top of the buttocks).

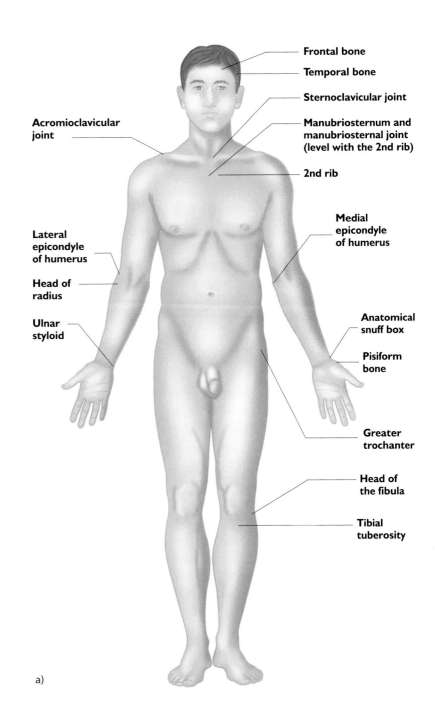

a)

Figure 92 (a–c): Bony landmarks.

b)

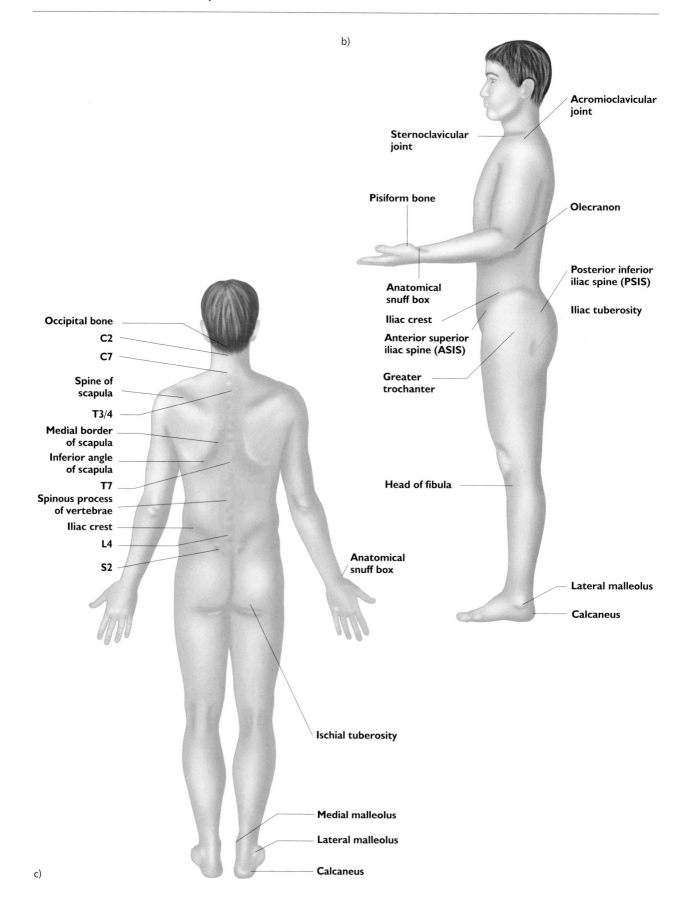

Acromioclavicular joint

Sternoclavicular joint

Pisiform bone

Olecranon

Anatomical snuff box

Posterior inferior iliac spine (PSIS)

Iliac crest

Iliac tuberosity

Anterior superior iliac spine (ASIS)

Greater trochanter

Head of fibula

Occipital bone

C2

C7

Spine of scapula

T3/4

Medial border of scapula

Inferior angle of scapula

T7

Spinous process of vertebrae

Iliac crest

L4

S2

Anatomical snuff box

Ischial tuberosity

Lateral malleolus

Calcaneus

Medial malleolus

Lateral malleolus

Calcaneus

c)

Figure 92 (a–c): Bony landmarks.

Joints

6

With the exception of the hyoid bone in the neck, all other bones form a joint with at least one other bone. Joints are also called *articulations*.

Joints have two functions: to *hold the bones together*; and to *give the rigid skeleton mobility*. When two bones meet, or articulate, there may or may not be movement depending on, (a) the amount of bonding material between the bones; (b) the nature of the material between the bones; (c) the shape of the bony surfaces; (d) the amount of the tension in the ligaments or muscles involved in the joint; (e) the position of the ligaments and muscles.

PART ONE–Classification of Joints

Joints are classified in two ways: **functionally** and **structurally**.

Functionally
The functional classification of joints focuses on the amount of movement allowed by the joint.

Immovable Joints (Synarthrotic)
These joints are found mostly in the axial skeleton, where joint stability and firmness is important for the protection of the internal organs.

Slightly Movable Joints (Amphiarthrotic)
Like immovable joints, and for the same reason, these joints are also found mainly in the axial skeleton.

Freely Movable Joints (Diarthrotic)
These joints predominate in the limbs, where a greater range of movement is required.

Structurally

Fibrous Joints
In fibrous joints, fibrous tissue joins the bones. As such, no joint cavity is present. Generally these joints have little or no movement, i.e. they are *synarthrotic*. Fibrous joints are of three types; *sutures, syndesmoses,* and *gomphoses*.

1. Sutures
The only examples of fibrous sutures are the sutures of the skull, where the irregular edges of the bones interlock and are bound tightly together by connective tissue fibres, allowing no active movement. Layers of periosteum on the inner and outer layers of the adjoining bones bridge the gap between the bones and form the main bonding factor. Between the adjoining joint surfaces there is a layer of vascular fibrous tissue that also helps unite the bones. This vascular fibrous tissue, along with the two layers of periosteum, is collectively called the *sutural ligament*. The fibrous tissue becomes ossified during adulthood by a process that occurs first at the deep aspect of the suture, progressively extending to the superficial part. This ossifying process is referred to as *synostosis*.

2. Syndesmoses
A syndesmosis is a fibrous joint where the uniting fibrous tissue forms an *interosseous membrane* or *ligament*; i.e. a band of fibrous tissue that allows little movement, situated between the radius and ulna and between the tibia and fibula.

3. Gomphoses
A gomphosis refers to a fibrous joint in which a peg is embedded into a socket. The only examples of such joints in humans consist of the teeth fixed into their sockets.

a)

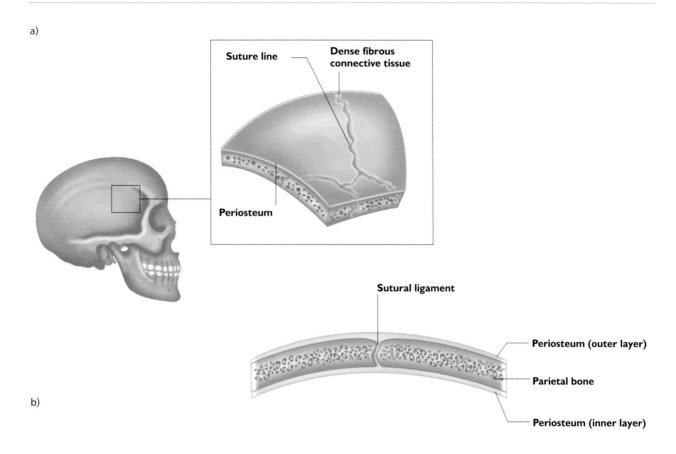

Suture line

Dense fibrous connective tissue

Periosteum

Sutural ligament

Periosteum (outer layer)

Parietal bone

Periosteum (inner layer)

b)

Figure 93: a) Position of a suture; b) vertical section through a suture.

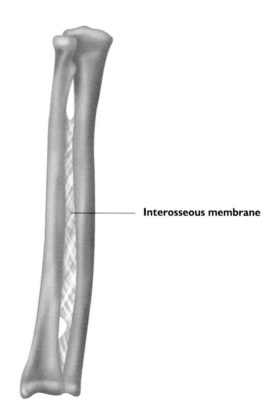

Interosseous membrane

Figure 94: The interosseous membrane between the radius and ulna.

Cartilaginous Joints

In cartilaginous joints, the bones are connected by a continuous plate of hyaline cartilage or a fibrocartilage disc. Again, no joint cavity is present. They can be either immovable (synchondrosis) or slightly movable (symphysis). The slightly moveable joints are the more common.

Synchondroses

Examples of cartilaginous joints that are immovable are the epiphyseal plates of growing long bones. These plates are made of hyaline cartilage that ossifies in young adults (*see* p.35). Thus, the place where a joint is united by such a plate is known as a *synchondrosis*. Another example of such a joint that eventually ossifies is the joint between the first rib and the manubrium of the sternum.

Symphyses

Examples of slightly movable cartilaginous joints are the pubic symphysis of the pelvic girdle, and the intervertebral joints of the spinal column. In both cases the articular surfaces of the bones are covered with hyaline cartilage that is in turn fused to a 'pad' of fibrocartilage (fibrocartilage is compressible and resilient, and acts as a shock absorber).

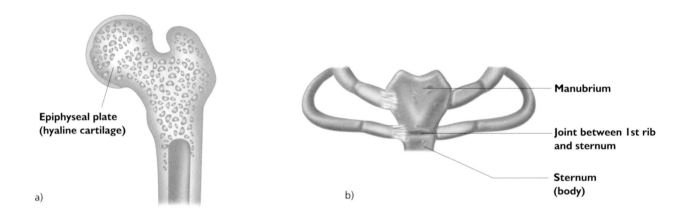

Figure 95: Cartilaginous immovable (synchondroses) joints (anterior view); a) the epiphyseal plate in a growing long bone, b) the sternocostal joint between manubrium and first rib.

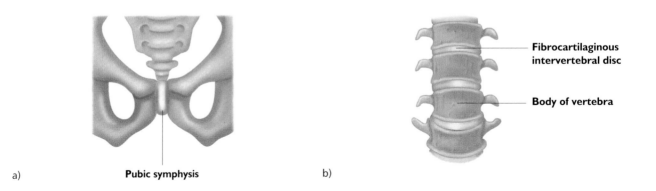

Figure 96: Cartilaginous slightly moveable (amphiarthrotic / symphysis) joints (anterior view); a) pubic symphysis of the pelvic girdle, b) intervertebral joints.

Synovial Joints

Synovial joints possess a joint cavity that contains *synovial fluid*. They are freely movable, *diarthrotic*, joints. Synovial joints have a number of distinguishing features:

Articular cartilage (or *hyaline cartilage*): covers the ends of the bones that form the joint.

A joint cavity: this cavity is more a potential space than a real one, because it is filled with lubricating *synovial fluid*. The joint cavity is enclosed by a double-layered 'sleeve' or capsule known as the *articular capsule*.

The external layer of the articular capsule is known as the *capsular ligament*. It is a tough, flexible, fibrous connective tissue that is continuous with the periostea of the articulating bones. The internal layer, or *synovial membrane*, is a smooth membrane made of loose connective tissue that lines the capsule and all internal joint surfaces other than those covered in hyaline cartilage.

Synovial fluid: a slippery fluid that occupies the free spaces within the joint capsule. Synovial fluid is also found within the articular cartilage and provides a film that reduces friction between the cartilages. When a joint is compressed by movement the fluid is forced out of the cartilage; when pressure is relieved the fluid rushes back into the articular cartilage. Synovial fluid nourishes the cartilage, which is *avascular* (contains no blood vessels); it also contains *phagocytic cells* (cells that eat dead matter) that rid the joint cavity of microbes or cellular waste. The amount of synovial fluid varies in different joints, but is always sufficient to form a thin film to reduce friction. During injury to the joint extra fluid is produced and creates the characteristic swelling of the joint. The synovial membrane later reabsorbs this extra fluid.

Collateral or accessory ligaments: synovial joints are reinforced and strengthened by a number of ligaments. These ligaments are either **capsular**, i.e. thickened parts of the fibrous capsule itself, or independent **collateral** ligaments that are distinct from the capsule. Ligaments always bind *bone to bone* and according to their position and quantity around the joint, they will restrict movement in certain directions, and prevent unwanted movement. As a general rule, the more ligaments a joint has, the stronger it is.

Bursae (sing. *bursa*) are fluid-filled sacs often found cushioning the joint. They are lined by synovial membrane and contain synovial fluid. They are found between tendons and bone, ligament and bone, or muscle and bone, and reduce friction by acting as a cushion.

Tendon sheaths are also frequently found in close proximity to synovial joints. They have the same structure as a bursa, and wrap themselves around tendons subject to friction, in order to protect them.

Articular discs (menisci) are present in some synovial joints. They act as shock absorbers (similar to the fibrocartilagenous disc in the pubic symphysis). For example, in the knee joint, two crescent-shaped fibrocartilage discs called the *medial* and *lateral menisci* lie between the medial and lateral condyles of the femur and the medial and lateral condyles of the tibia.

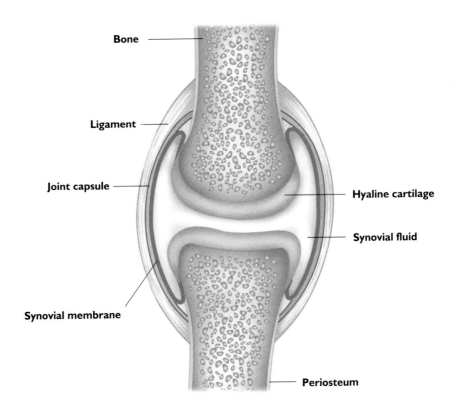

Bone

Ligament

Joint capsule

Hyaline cartilage

Synovial fluid

Synovial membrane

Periosteum

Figure 97: A typical synovial joint.

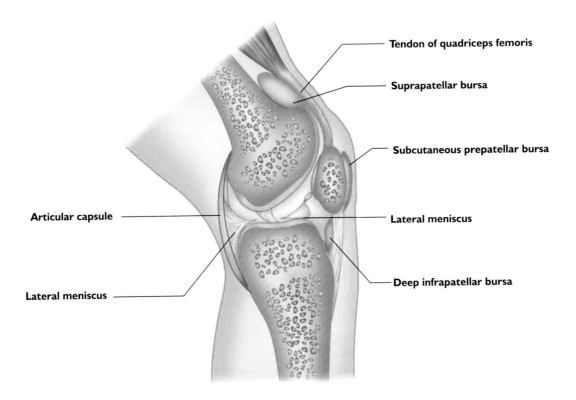

Tendon of quadriceps femoris

Suprapatellar bursa

Subcutaneous prepatellar bursa

Articular capsule

Lateral meniscus

Lateral meniscus

Deep infrapatellar bursa

Figure 98: Shock absorbing and friction-reducing structures of a synovial joint.

The Seven Types of Synovial Joints

Plane or Gliding

In gliding joints, movement occurs when two, generally flat or slightly curved, surfaces glide across one another. Examples: the acromioclavicular joint; the joints between the carpal bones in the wrist, or the tarsal bones in the ankle; the facet joints between the vertebrae; the sacroiliac joint.

Hinge

In hinge joints, movement occurs around only one axis; a transverse one – as in the hinge of the lid of a box. A protrusion of one bone fits into a concave or cylindrical articular surface of another, permitting flexion and extension. Examples: the interphalangeal joints, the elbow, and the knee.

Pivot

In pivot joints, movement takes place around a vertical axis, like the hinge of a gate. A more or less cylindrical articular surface of bone protrudes into and rotates within a ring formed by bone or ligament. Examples: the dens of the axis protrudes through the hole in the atlas, allowing the rotation of the head from side to side. Also, the joint between the radius and the ulna at the elbow allows the round head of the radius to rotate within a 'ring' of ligament that is secured to the ulna.

Ball and Socket

Ball and socket joints consist of a 'ball' formed by the spherical or hemispherical head of one bone that rotates within the concave socket of another, allowing flexion, extension, adduction, abduction, circumduction, and rotation. Thus, they are multiaxial and allow the greatest range of movement of all joints. Examples: the shoulder and the hip joints.

Condyloid

In common with ball and socket joints, condyloid joints have a spherical articular surface that fits into a matching concavity. Also, like ball and socket joints, condyloid joints permit flexion, extension, abduction, adduction, and circumduction. However, the disposition of surrounding ligaments and muscles prevent active rotation around a vertical axis. Examples: the metacarpophalangeal joints of the fingers (but not the thumb).

Saddle

Saddle joints are similar to condyloid joints, except that both articulating surfaces have convex and concave areas, and so resemble two 'saddles' that join them together by accommodating each others convex to concave surfaces. Saddle joints allow even more movement than condyloid joints, for example, allowing the 'opposition' of the thumb to the fingers. Example: the carpometacarpal joint of the thumb.

Ellipsoid

An ellipsoid joint is effectively similar to a ball and socket joint, but the articular surfaces are ellipsoid instead of spherical. Movements as for ball and socket joints, with the exception of rotation, which is prevented by the shape of the ellipsoid surfaces. Example: the radio-carpal joint.

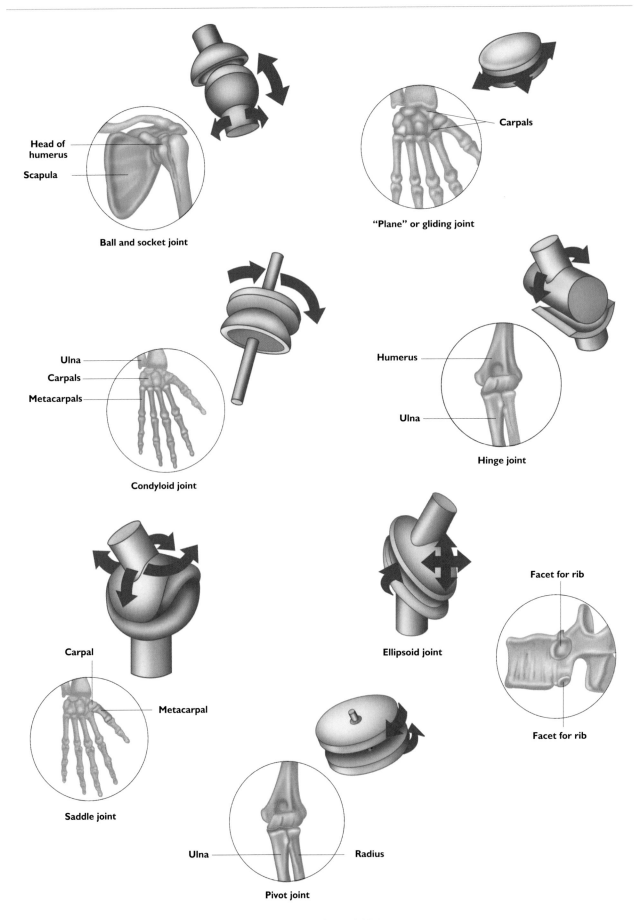

Figure 99: Types of synovial joints.

PART TWO–Features of Specific Joints

This section presents a relatively in-depth examination of four synovial joints: i.e. the shoulder, elbow, hip and knee joints. The other joints, of all classifications, are presented in less detail, with minimal accompanying text, but with fully labelled illustrations.

Notes About Synovial Joints:

- Some tendons run partly within the joint and are therefore intracapsular.

- The fibres of many ligaments are largely integrated with those of the capsule and the delineation between capsule and ligament is sometimes unclear. Therefore, only the main ligaments are mentioned.

- Ligaments are termed intracapsular (or intra-articular) when inside the joint cavity, and extracapsular (or extra-articular) when outside the capsule.

- Many ligaments of the knee joint are modified extensions or expansions of muscle tendons, but are classed as ligaments to differentiate them from the more regular stabilizing tendons, such as the patellar ligament from the quadriceps.

- Most synovial joints have various bursae in their vicinity, as shown in the illustrations pertaining to each joint.

Joints of the Head and Vertebral Column

Temporomandibular Joint

Type of Joint
Synovial hinge joint, plus a plane joint.

Articulation
The head of the mandible articulates with the mandibular fossa and the articular tubercle of the temporal bone. A fibrous disc separates the articular surfaces and moulds itself upon them when the joint moves.

Movements
This is the only moveable joint in the head. Movement can occur in all three planes; upwards and downwards, backwards and forwards, and from side to side. A gliding action occurs superior to the disc. A hinge action occurs inferior to the disc.

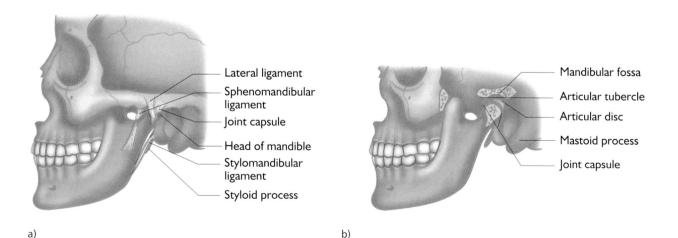

a)   b)

Figure 100: The temporomandibular joint; a) lateral view, b) lateral view.

Atlanto-occipital Joint

Type of Joint
The articulations of the two sides act together functionally as a synovial ellipsoid joint.

Articulation
Between the occipital condyles and the superior articular facets of the atlas.

Movements
Flexion and extension (as in nodding the head). Lateral flexion.

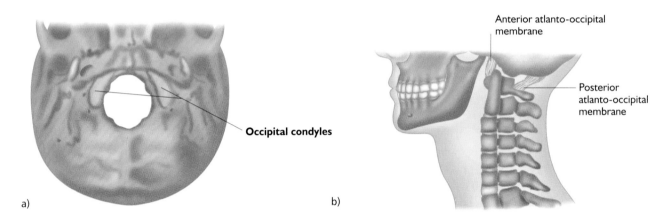

Figure 101: The atlanto-occipital joint; a) inferior view, b) lateral view.

Atlanto-axial Joint

Type of Joint
Lateral atlanto-axial joint: Synovial plane.
Median (sagittal) atlanto-axial joint: Synovial pivot.

Articulation
Lateral atlanto-axial joint: Between the opposed articular processes of the atlas and axis.
Median (sagittal) atlanto-axial joint: Between the dens of the axis and the anterior arch of the atlas, and with the transverse ligament.

Movements
Rotation of the head around a vertical axis (the skull and the atlas moving as one).

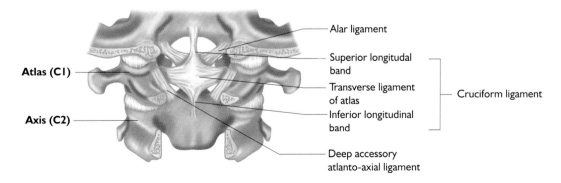

Figure 102: The atlanto-axial joint (posterior view).

Joints Between Vertebral Bodies

Type of Joint
Cartilaginous symphysis (slightly moveable).

Articulation
Between adjacent surfaces of vertebral bodies united by a fibrocartilaginous intervertebral disc.

Movements
Only slight movement occurs between any two successive vertebrae, but there is considerable movement throughout the column as a whole.

Cervical region: Flexion, extension, lateral flexion with rotation (i.e. lateral flexion cannot occur without an element of rotation and vice versa).
Thoracic region: Rotation, always associated with an element of lateral flexion, and vice versa. Only extremely slight flexion and extension can occur (limited by presence of ribs and sternum).
Lumbar region: Flexion, extension. Only extremely slight rotation can occur (restricted by the angle of the articular processes).

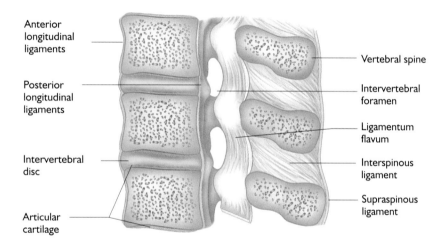

Figure 103a: Sagittal section through 2nd to 4th lumbar vertebrae.

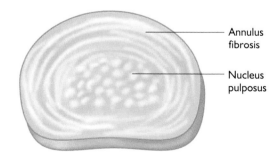

Figure 103b: Transverse section of a lumbar intervertebral disc.

Joints Between Vertebral Arches

Type of Joint
Synovial plane.

Articulation
Between opposed articular processes of adjacent vertebrae, to unite adjacent vertebral arches.

Movements
Only slight movement occurs between any two successive vertebrae, but there is considerable movement throughout the column as a whole.

Cervical region: Flexion, extension, lateral flexion with rotation (i.e. lateral flexion cannot occur without an element of rotation and vice versa).
Thoracic region: Rotation, always associated with an element of lateral flexion, and vice versa. Only extremely slight flexion and extension can occur (limited by presence of ribs and sternum).
Lumbar region: Flexion, extension. Only extremely slight rotation can occur (restricted by the angle of the articular processes).

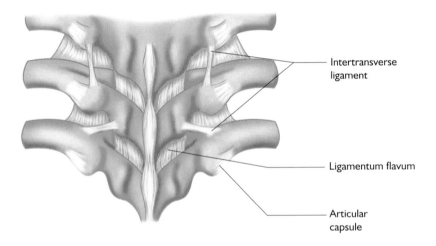

Intertransverse ligament

Ligamentum flavum

Articular capsule

Figure 104: A typical vertebral arch joint (posterior view).

Joints of the Ribs and Sternum

Costovertebral Joints

Type of Joint
Joints of the heads of ribs (capitular joints): Synovial plane.
Costotransverse joints: Synovial plane.

Articulation
Joints of the heads of ribs (capitular joints): Superior and inferior articular facets on the head of each typical rib articulate with the facets on two adjacent vertebral bodies (i.e. the rib's head sits between two vertebral bodies, and also against a shallow depression on the intervertebral disc).
Costotransverse joints: The tubercle of each typical rib articulates with the transverse process of the lower of the two vertebrae to which its head is joined (but ligaments attach it to the transverse processes of both vertebrae).

NOTE: The first rib and last two or three ribs have atypical vertebral connections, because the head of these ribs have only one facet, not two; and therefore articulate with one vertebral body rather than two. The tubercles of the lowest ribs do not form synovial joints with the transverse processes.

Movements
The capitular and costotransverse joints of each rib together form a hinge, causing the anterior part of the rib to be raised (with some lateral 'expansion') during inspiration, and lowered (with some medial 'contraction') during expiration. This effectively increases and decreases the anteroposterior and transverse diameters of the thorax with each in-breath and out-breath.

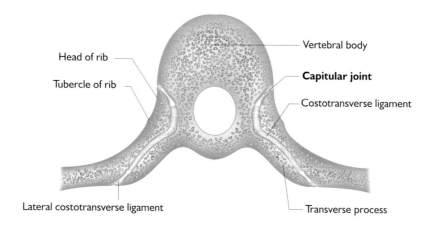

Figure 105a: Transverse section through a typical costovertebral joint.

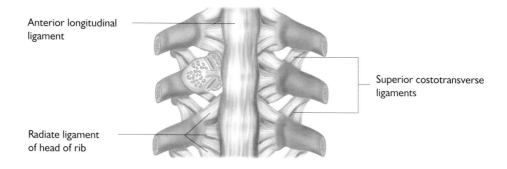

Figure 105b: The costovertebral joint (anterior view).

Sternocostal Joints

The hyaline cartilage that is continuous with the anterior end of each rib is called the *costal cartilage*.

Type of Joint
First rib: Cartilaginous immovable (synchondrosis).
Ribs 2–7: Simple synovial plane.
Ribs 8–10: Simple synovial plane articulations at interchondral joints.

Articulation
First rib: Via costal cartilage to the body of the sternum.
Ribs 2–7: Via costal cartilages to facets on the side of the body of the sternum. The joint cavities are divided in two by an intra-articular ligament (until cavities disappear in old age).
Ribs 8–10: Their costal cartilages unite with the costal cartilage of rib 7.
Ribs 11–12: Do not articulate anteriorly, but end freely in the muscles of the flank. They are therefore called floating ribs.

Movements
Enables expansion and contraction of the ribcage, as described under costovertebral joints (*see* p.93).

Sternal Joints

Type of Joint
Manubriosternal joint: Similar in appearance to a cartilaginous symphysis (slightly moveable) joint.
Xiphisternal joint: Cartilaginous immovable (synchondrosis). Usually becomes ossified in old age.

Articulation
Manubriosternal joint: Between the manubrium and body of the sternum, adjacent to the second costal cartilage.
Xiphisternal joint: Between the body of the sternum and the xiphoid process. This joint marks the inferior extent of the thoracic cavity.

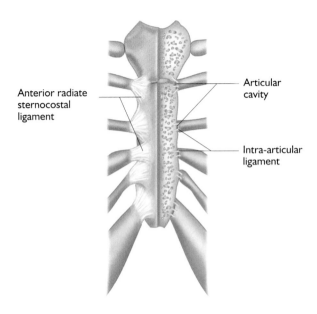

Anterior radiate sternocostal ligament

Articular cavity

Intra-articular ligament

Figure 106: The sternocostal joint (anterior view).

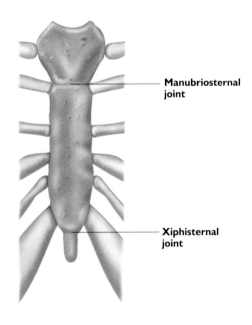

Manubriosternal joint

Xiphisternal joint

Figure 107: The sternal joints (anterior view).

Joints of the Shoulder Girdle and Upper Limb

Sternoclavicular Joint

Type of Joint
Functionally a synovial ball and socket (but unlike most articular surfaces, the articular cartilage is fibrocartilage rather than hyaline cartilage).

Articulation
Between the sternal (medial) end of the clavicle, the clavicular notch of the manubrium and the costal cartilage of the first rib.

NOTE: A fibrocartilage articular disc separates the joint space into two separate synovial cavities.

Movement
Like other ball and socket joints, movement occurs in all planes, but anteroposterior movement and rotation is slightly restricted. It is involved in the collective movements of the shoulder girdle.

Acromioclavicular Joint

Type of Joint
Synovial plane.

Articulation
Between the lateral end of the clavicle and the medial border of the acromion of the scapula.

NOTE: A fibrocartilage articular disc partially divides the articular cavity, although it is sometimes absent.

Movement
It is involved in the collective movements of the shoulder girdle, enabling the scapula to change its position in relation to the clavicle.

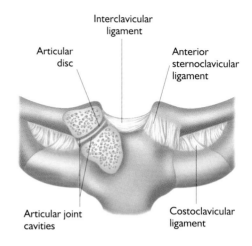

Figure 108: The sternoclavicular joint (anterior view). Note: Posterior aspect of joint has a posterior sternoclavicular ligament similar, but weaker, than anterior sternoclavicular ligament.

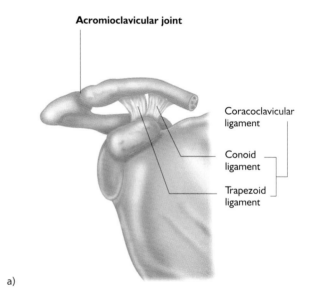

a)

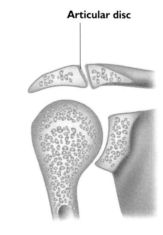

b)

Figure 109: The acromioclavicular joint; a) anterior view, b) coronal view.

Shoulder (Glenohumeral) Joint

Type of Joint
Synovial ball and socket.

Articulation
The head of the humerus articulates with the shallow pear shaped glenoid cavity (fossa) of the scapula. This articulation is inherently unstable due to the glenoid cavity being only approximately one-third the size of the humeral head, although it is slightly deepened by a rim of fibrocartilage called the *glenoid labrum* or *labrum glenoidale* (triangular in cross-section). The shoulder joint is the most freely moving joint of the body; precisely because stability has been sacrificed to enable maximum range of movement.

Articular Capsule
Extends from the margin of the glenoid cavity (including part of the labrum), to the anatomical neck of the humerus. The thin capsule is very loose, thus enabling maximum movement of the joint. When the arm is by the side, the lower part of the capsule hangs in a loose fold, which becomes progressively more taut as the arm is abducted; increasingly so if the arm continues into elevation. The capsule contributes very little to the stability of the joint. Joint stability is largely supplied by the surrounding muscles, whose attachments are intimately related to the capsule.

Ligaments
Transverse humeral ligament: Spans the gap between the humeral tubercles. It holds the long head of the biceps brachii in the intertubular sulcus as it leaves the joint.
Glenohumeral ligament: Three slightly thickened bands of longitudinal fibres on the internal surface of the anterior part of the capsule. May be absent.
Coracohumeral ligament: Extends from the coracoid process of the scapula to the upper part of the anatomical neck of the humerus. It greatly reinforces the capsule superiorly and slightly anteriorly.
Coraco-acromial ligament: This ligament is totally unconnected to the articular capsule. It forms a shelf above the joint, running between the coracoid process and the acromion process of the scapula.

Various bursae are associated with the shoulder joint. The most important is the subacromial bursa that separates the coraco-acromial ligament from the supraspinatus tendon located above the shoulder joint.

Stabilizing Tendons
Long head of biceps brachii tendon: Runs from the superior aspect of the glenoid labrum to enter and travel within the joint cavity, thus travelling within the articular capsule (hence it is covered with a sheath of synovial membrane). On leaving the cavity, it enters the intertubular groove of the humerus. Its location secures the head of the humerus tightly against the glenoid cavity, thereby acting as a steadying influence during movements of the shoulder joint.
Rotator cuff tendons: The four rotator cuff tendons (supraspinatus, infraspinatus, teres minor, and subscapularis: *see* pp.266–269) encircle the joint and fuse with the articular capsule. Consequently, the rotator cuff muscles or tendons are prone to injury if the joint is vigorously circumducted, as in throwing a ball.

NOTE: Because overall, the reinforcements of the shoulder joint are weakest inferiorly, the humerus is more prone to dislocate downwards.

Movements
Flexion, extension, abduction, adduction, medial and lateral rotation, circumduction, plus elevation through flexion and abduction (*see* p.20–22, 25).

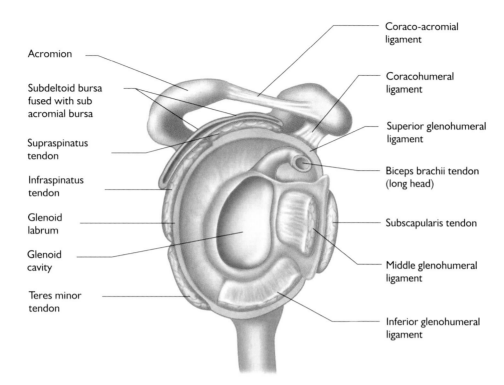

Coraco-acromial ligament

Acromion

Coracohumeral ligament

Subdeltoid bursa fused with sub acromial bursa

Superior glenohumeral ligament

Supraspinatus tendon

Biceps brachii tendon (long head)

Infraspinatus tendon

Subscapularis tendon

Glenoid labrum

Glenoid cavity

Middle glenohumeral ligament

Teres minor tendon

Inferior glenohumeral ligament

a)

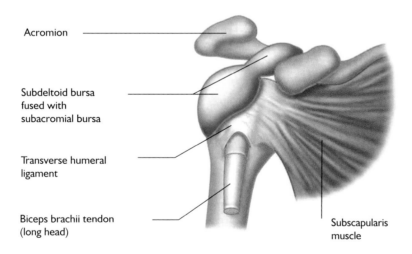

Acromion

Subdeltoid bursa fused with subacromial bursa

Transverse humeral ligament

Biceps brachii tendon (long head)

Subscapularis muscle

b)

Figure 110: The shoulder joint; a) right arm, lateral view, b) right arm, anterior view (cut).

Elbow Joint

Type of Joint
Synovial hinge (ginglymus).

Articulation
Upper surface of the head of the radius articulates with the capitulum of the humerus. The trochlear notch of the ulna articulates with the trochlea of the humerus (which constitutes the 'hinge' mechanism and the main stabilizing factor).

Articular Capsule
The relatively loose articular capsule extends from the coronoid and olecranon fossae of the humerus to the coronoid and olecranon processes of the ulna, and to the annular ligament enclosing the head of the radius. The capsule is thin anteriorly and posteriorly to allow flexion and extension, but is strengthened on each side by collateral ligaments.

Ligaments
Ulnar (medial) collateral ligament: Three strong bands reinforcing the medial side of the capsule.
Radial (lateral) collateral ligament: A strong triangular ligament reinforcing the lateral side of the capsule.

Stabilizing Tendons
The tendons of the biceps brachii, triceps brachii, brachialis, plus many muscles located on the forearm: These tendons cross the elbow joint and provide extra security.

Movements
Flexion and extension only.

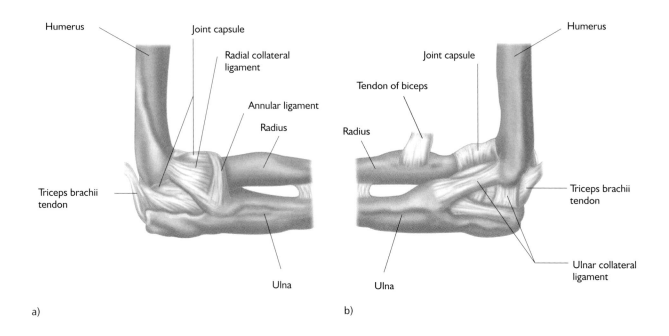

Humerus

Joint capsule

Radial collateral
ligament

Annular ligament

Radius

Triceps brachii
tendon

Ulna

a)

Humerus

Joint capsule

Tendon of biceps

Radius

Triceps brachii
tendon

Ulnar collateral
ligament

Ulna

b)

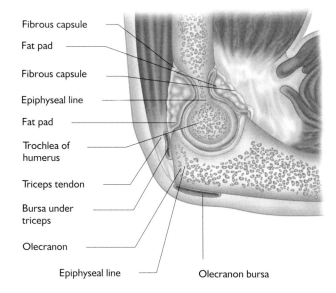

Fibrous capsule

Fat pad

Fibrous capsule

Epiphyseal line

Fat pad

Trochlea of
humerus

Triceps tendon

Bursa under
triceps

Olecranon

Epiphyseal line

Olecranon bursa

c)

Figure 111: The elbow joint; a) right arm, lateral view, b) right arm, medial view, c) right arm, mid-sagittal view.

Proximal Radio-ulnar Joint

Type of Joint
Synovial pivot.

Articulation
The disc shaped head of the radius rotates within a ring formed by the radial notch on the ulna and the annular ligament of the radius.

NOTE: The synovial cavity of this joint is continuous with that of the elbow joint.

Movements
Pronation and supination of the forearm.

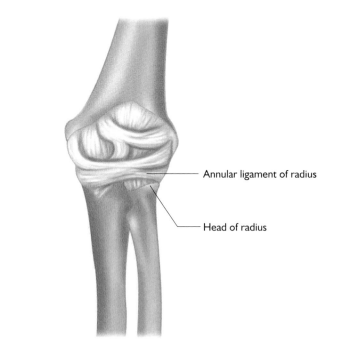

Annular ligament of radius

Head of radius

a)

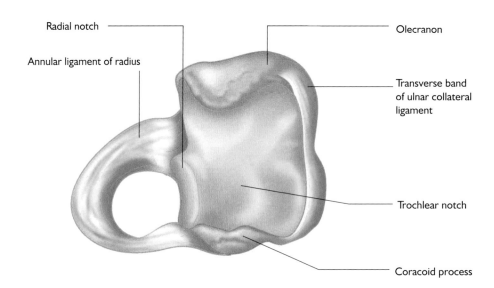

Radial notch

Olecranon

Annular ligament of radius

Transverse band of ulnar collateral ligament

Trochlear notch

b)

Coracoid process

Figure 112: The proximal (superior) radio-ulnar joint; a) left arm, anterior view, b) left arm, superior view.

Distal Radio-ulnar Joint

Type of Joint
Synovial pivot.

Articulation
Between the head of the ulna and the ulnar notch of the radius.

NOTE: A fibrocartilage articular disc unites the styloid process of the ulna and the medial side of the distal radius.

Movements
Pronation and supination of the forearm.

Intermediate Radio-ulnar Joint

Type of Joint
Syndesmosis.

Articulation
Connects the interosseous border of the radius with the interosseous border of the ulna, via the interosseous membrane. Also, a slender fibrous band called the *oblique cord* connects the ulnar tuberosity to the proximal end of the shaft of the radius.

Function
Increases the surface of origin of the deep forearm muscles; helps bind the radius and ulna together; and transmits to the ulna any force passing upwards from the hand along the radius.

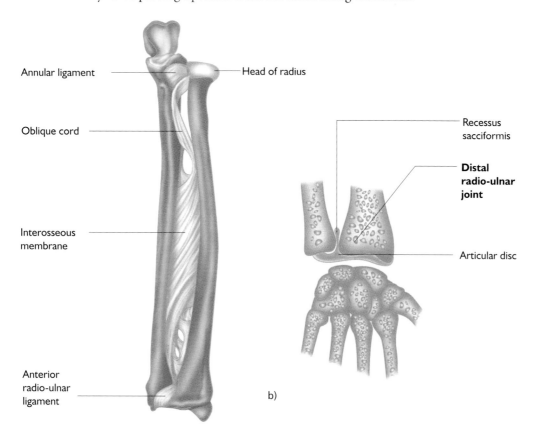

Figure 113: The distal and intermediate radio-ulnar joints; a) left arm / hand, anterior view, b) left arm / hand, coronal view.

Radio-carpal Joint (Wrist Joint)

Type of Joint
Synovial ellipsoid.

Articulation
The distal surface of the radius and the articular disc (the same disc as described with the distal radio-ulnar joint, *see* p.101) articulates with the proximal row of carpals, which are the scaphoid, lunate and triquetral (triquetrum).

Movements
Movements are in combination with the intercarpal joints: flexion, extension, adduction (ulnar deviation), abduction (radial deviation) and circumduction.

Intercarpal Joints

Type of Joint
A series of synovial plane joints.

Articulation
This joint has articulations between the two carpal rows (midcarpal joint), plus articulations between each bone of the proximal carpal row and of the distal carpal row.

Movements
Movements are in combination with the radio-carpal joint: flexion, extension, adduction (ulnar deviation), abduction (radial deviation) and circumduction.

Carpometacarpal Joint of the Thumb

Type of Joint
Synovial saddle joint.

Articulation
Between the trapezium and the base of the first metacarpal bone (the thumb).

Movements
Flexion, extension, abduction and adduction. At the extreme range of flexion, the first metacarpal medially rotates so that the palmar surface of the thumb becomes opposed to the pads of the fingers. Conversely, slight lateral rotation occurs when the thumb approaches full extension. Combining these movements create approximate circumduction of the thumb.

Common Carpometacarpal Joint

Type of Joint
Synovial plane.

Articulation
Between the distal row of carpal bones and the bases of the medial four metacarpal bones of the hand.

Movements
Very little movement is possible. However, the articulation at the fifth metacarpal with the hamate is a flattened saddle joint, allowing slight opposition of the little finger across the palm.

Intermetacarpal Joints

Type of Joint
Synovial plane.

Articulation
Between adjacent sides of the bases of metacarpal bones 2–5.

Movements
Limited movement between adjacent metacarpals.

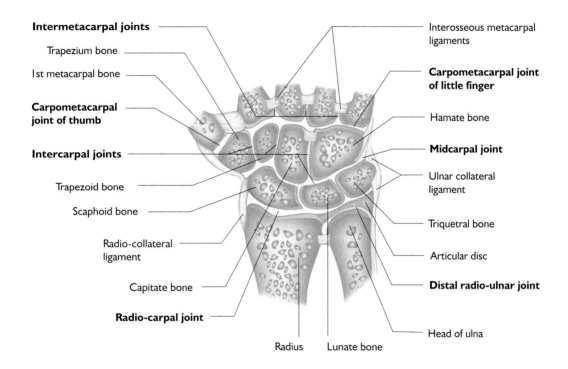

Figure 114: The radio-carpal (wrist), intercarpal, carpometacarpal and intermetacarpal joints (coronal view).

Metacarpophalangeal Joints

Type of Joint
Synovial condyloid.

Articulation
Between the head of a metacarpal and the base of a proximal phalanx.

NOTE: The capsule is deficient on the dorsal aspect, where it is replaced by an expansion of the long extensor tendon.

Movements
Flexion and extension. Abduction and adduction (possible only in extension, but with very little movement at the thumb). Combined movements may produce circumduction.

Interphalangeal Joints

Type of Joint
Synovial hinge.

Articulation
Between the proximal and middle phalanges (proximal interphalangeal joint, abbreviated PIP), or the middle and distal phalanges (distal interphalangeal joint, abbreviated DIP).

Movements
Flexion and extension.

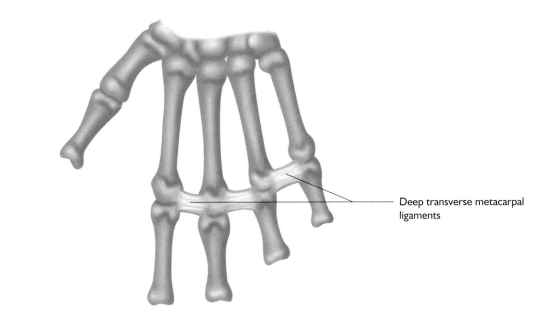

Deep transverse metacarpal ligaments

a)

Metacarpophalangeal joint

Joint capsule

Metacarpal bone

Collateral ligament

Proximal interphalangeal joint

Distal interphalangeal joint

Palmar ligament (the thumb has two sesamoid bones embedded; the index finger and occasionally the little finger, have one)

Proximal   Middle   Distal

Phalanges

b)

Figure 115: The metacarpophalangeal and interphalangeal joints; a) anterior view, b) medial view.

Joints of the Pelvic Girdle and Lower Limb

Lumbosacral and Sacrococcygeal Joints

Type of Joint
Both joints: cartilaginous symphysis (slightly moveable).

Articulation
Lumbosacral: Between the fifth lumbar vertebra (L5) and the body of the first sacral segment (S1). This joint has the same features as other typical intervertebral joints, with the addition of the iliolumbar ligament.
Sacrococcygeal: Between the last sacral and first coccygeal segments. It is reinforced all round by the sacrococcygeal ligaments.

NOTE: Both joints contain a fibrocartilaginous intervertebral disc.

Movements
The lumbosacral joint contributes to the collective movements of the lumbar vertebral joints. The sacrococcygeal joint has very little functional movement, and us often partially or fully obliterated in old age.

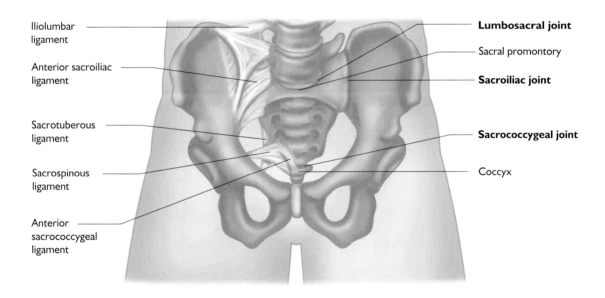

Iliolumbar ligament
Anterior sacroiliac ligament
Sacrotuberous ligament
Sacrospinous ligament
Anterior sacrococcygeal ligament

Lumbosacral joint
Sacral promontory
Sacroiliac joint
Sacrococcygeal joint
Coccyx

Figure 116: The lumbosacral, sacroiliac and sacrococcygeal joints (anterior view).

Sacroiliac Joint

Type of Joint
A synovial joint with pronounced irregular depressions and tubercles on the articular surfaces.

NOTE: The articular surface of the sacrum is hyaline cartilage, but that of the ilium is usually of the fibrous type.

Articulation
Between the auricular surfaces on the sacrum and the iliac bone.

Movements
Very limited movements occur because of the irregular joint surfaces and the strong sacroiliac ligaments.

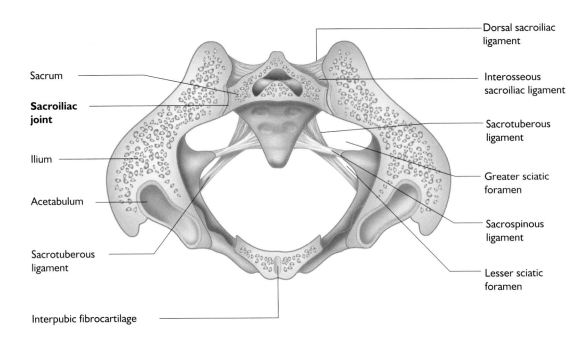

Sacrum

Sacroiliac joint

Ilium

Acetabulum

Sacrotuberous ligament

Interpubic fibrocartilage

Dorsal sacroiliac ligament

Interosseous sacroiliac ligament

Sacrotuberous ligament

Greater sciatic foramen

Sacrospinous ligament

Lesser sciatic foramen

Figure 117: Transverse section of pelvis.

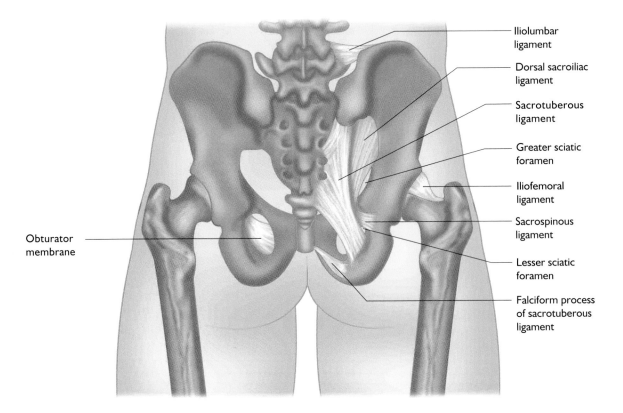

Obturator membrane

Iliolumbar ligament

Dorsal sacroiliac ligament

Sacrotuberous ligament

Greater sciatic foramen

Iliofemoral ligament

Sacrospinous ligament

Lesser sciatic foramen

Falciform process of sacrotuberous ligament

Figure 118: Pelvic ligaments (posterior view).

Pubic Symphysis

Type of Joint
Cartilaginous symphysis (slightly moveable).

Articulation
The midline joint between the superior rami of the pubic bones.

NOTE: The joint contains a fibrocartilaginous interpubic disc with a slit-like cavity, which in women, can develop into a large cavity.

Movements
No significant movement occurs other than some separation of the pubic bones in women during pregnancy and childbirth.

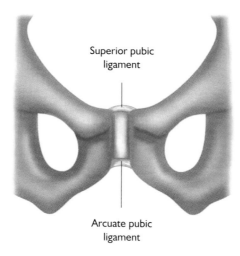

Figure 119: Pubic symphysis (anterior view).

Hip Joint

Type of Joint
Synovial ball and socket.

Articulation
The spherical head of the femur articulates with the cup-like acetabulum of the coxal (hip) bone. The depth of the acetabulum is enhanced by a circular rim of fibrocartilage called the *acetabular labrum* or *labrum acetabulare*, which grasps the femoral head. Unlike the articulation of the shoulder joint, the hip articulation fits securely together.

Articular Capsule
Extends from the rim of the acetabulum to the neck of the femur. It is very strong and tense in extension, which contrasts to the thin and lax capsule of the shoulder joint.

Ligaments
Iliofemoral ligament: A thick and strong triangular band situated anteriorly.
Pubofemoral ligament: A triangular thickening of the inferior aspect of the capsule.
Ischiofemoral ligament: A spiral ligament situated posteriorly.

These three ligaments are arranged so that when a person stands up (i.e. hip joint moves from flexion to extension), the head of the femur is 'screwed' into the acetabulum, and held firmly in position.

Ligament of the head of the femur: Also called the ligamentum teres or the capitate ligament, this flat intracapsular ligament runs from the femoral head to the lower lip of the acetabulum. It contains an artery that is a source of blood for the head of the femur. It is slack during most hip movements and therefore does not contribute to the joint's stability.

Stabilizing Tendons
This joint is inherently stable by virtue of its structure and ligaments. All surrounding muscles and tendons contribute to its stability, but in a very minor capacity compared to those of the shoulder joint.

Movements
Flexion, extension, abduction, adduction, medial and lateral rotation, circumduction (limited, compared to the shoulder joint).

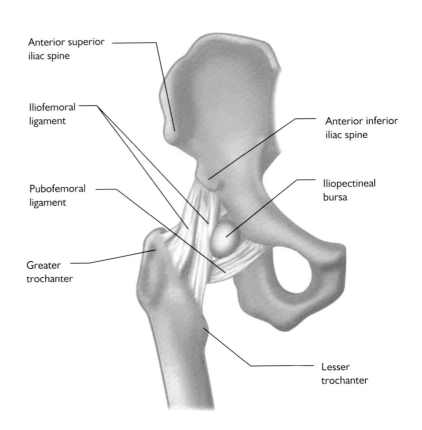

Anterior superior iliac spine

Iliofemoral ligament

Pubofemoral ligament

Greater trochanter

Anterior inferior iliac spine

Iliopectineal bursa

Lesser trochanter

a)

Figure 120: The hip joint, a) right leg, anterior view, b) right leg, posterior view, c) right leg, lateral view.

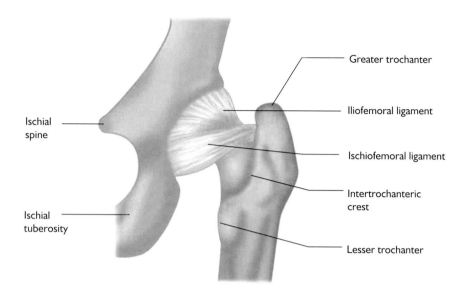

Greater trochanter

Iliofemoral ligament

Ischial
spine

Ischiofemoral ligament

Intertrochanteric
crest

Ischial
tuberosity

Lesser trochanter

b)

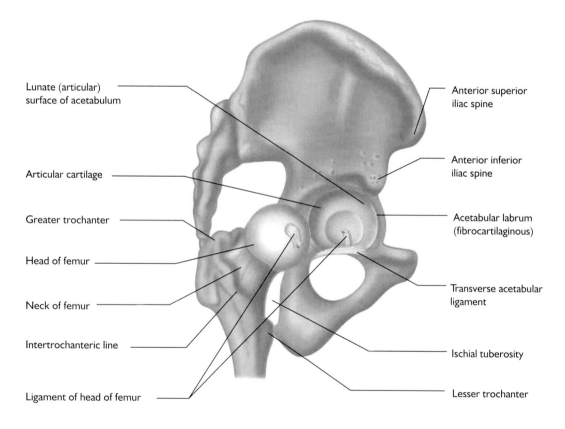

Lunate (articular)
surface of acetabulum

Anterior superior
iliac spine

Articular cartilage

Anterior inferior
iliac spine

Greater trochanter

Acetabular labrum
(fibrocartilaginous)

Head of femur

Neck of femur

Transverse acetabular
ligament

Intertrochanteric line

Ischial tuberosity

Lesser trochanter

Ligament of head of femur

c)

Figure 120 (cont.): The hip joint, a) right leg, anterior view, b) right leg, posterior view, c) right leg, lateral view.

Knee Joint

The knee joint is the largest and most complex joint in the body. Within its joint cavity it contains three articulations: the lateral and medial articulations of the *tibiofemoral joint*, and the *femoropatellar joint*.

Type of Joint

Tibiofemoral joint: Functionally a modified synovial hinge joint, but structurally a condyloid joint.
Femoropatellar joint: Synovial plane joint.

Articulation

Tibiofemoral joint: Condyles of the femur articulate with the condyles of the tibia; but with two C-shaped menisci or semilunar cartilages between the opposing articular surfaces.
Femoropatellar joint: Posterior surface of patella articulates with patellar surface at the lower end of the femur.

Articular Capsule

The knee is the only joint where the capsule only partially encloses the joint cavity. Instead, the true capsular fibres are integrated within a ligamentous sheath composed of muscle tendons or expansions from them, which collectively encapsulate the joint. True capsular fibres are located only at the sides and posterior of the joint.

Extracapsular (extra-articular) Ligaments

Tibial (medial) collateral ligament: A broad, flat band running from the medial epicondyle of the femur, downwards and forwards to the medial condyle of the tibial shaft. Some fibres are fused to the medial meniscus.
Fibular (lateral) collateral ligament: A round, cord-like ligament, fully detached from the thin lateral part of the capsule. It extends from the lateral epicondyle of the femur, downwards and backwards to the head of the fibula.
Oblique popliteal ligament: An expansion of the semimembranosus tendon, that passes upward and laterally over the posterior of the joint.
Arcuate popliteal ligament: Extends from the head of the fibula upwards and medially, spreading into the back of the capsule and to the lateral condyle of the femur; thus reinforcing the back of the joint.

Intracapsular (intra-articular) Ligaments and Menisci

Anterior cruciate ligament: Extends obliquely upwards, laterally and backwards from the anterior intercondylar area of the tibia to the medial surface of lateral femoral condyle. It prevents posterior displacement of the femur on the tibia, and also helps check hyperextension of the knee.
Posterior cruciate ligament: Passes upwards, medially and forwards from the posterior intercondylar area of the tibia to the lateral side of the medial femoral condyle. Thus it lies on the medial side of the weaker anterior cruciate. It prevents anterior displacement of the femur on the tibia.

The cruciate ligaments are within the joint capsule, but outside the joint cavity. Synovial membrane covers most of their surface.

Menisci: Between the femoral and tibial condyles are two crescent shaped fibrous wedges called menisci, that help compensate for the incongruence of the articular surfaces. They also help absorb shock transmitted to the knee joint. The menisci are attached only at their outer margins and are prone to tearing. The medial meniscus is also attached to the tibial collateral ligament, and is therefore much more firmly anchored than the lateral meniscus, which does not attach to the fibular collateral ligament.
Medial and lateral coronary ligaments: Capsular fibres that attach the menisci to the tibial condyles.
Transverse ligament of the knee: A fibrous band that joins the anterior parts of the menisci.

Stabilizing Tendons

Patellar ligament (ligamentum patellae): This strong ligament is actually the distal part of the quadriceps tendon. It runs from the patella (which is embedded within the tendon as a sesamoid bone – *see* p.39) to the tibial tuberosity. Other thinner bands called the *medial and lateral patellar retinacula* pass down the sides of the patella to attach to the front of each tibial condyle; effectively substituting for the capsule anteriorly.

The Atlas of Musculo-skeletal Anatomy   111

Joints

Tendon of semimembranosus: Helps reinforce the posterior of the knee joint.

The muscles surrounding the knee joint are particularly crucial as stabilizers.

Movements

Flexion, extension. Some rotation can occur when the knee is flexed. Also, as a result of the tightening of various ligaments (especially the cruciates) and tendons, slight medial rotation of the femur occurs upon the fixed tibia as the knee straightens into full extension. (When both the femur and tibia are not fixed, as in kicking, the tibia rotates laterally at the end of extension and medially at the beginning of flexion).

NOTE: The popliteus muscle 'unlocks' the extended knee joint prior to flexion, enabling flexion to occur (*see* popliteus – p.350).

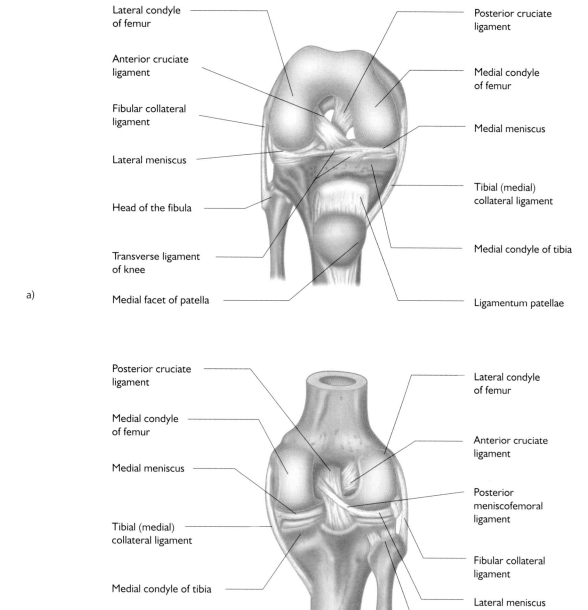

Figure 121: The knee joint; a) right leg, anterior view, b) right leg, posterior view, c) right leg, posterior view, d) right leg, mid-sagittal view.

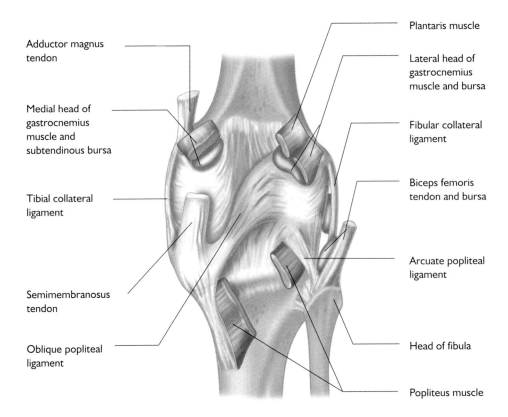

Adductor magnus tendon

Medial head of gastrocnemius muscle and subtendinous bursa

Tibial collateral ligament

Semimembranosus tendon

Oblique popliteal ligament

Plantaris muscle

Lateral head of gastrocnemius muscle and bursa

Fibular collateral ligament

Biceps femoris tendon and bursa

Arcuate popliteal ligament

Head of fibula

Popliteus muscle

c)

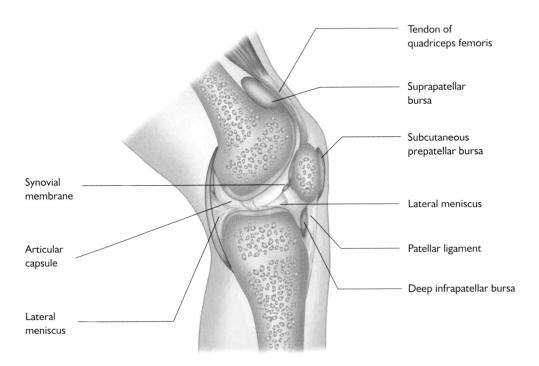

Tendon of quadriceps femoris

Suprapatellar bursa

Subcutaneous prepatellar bursa

Lateral meniscus

Patellar ligament

Deep infrapatellar bursa

Synovial membrane

Articular capsule

Lateral meniscus

d)

Figure 121 (cont.): The knee joint; a) right leg, anterior view, b) right leg, posterior view, c) right leg, posterior view, d) right leg, mid-sagittal view.

Proximal Tibiofibular Joint

Type of Joint
Synovial plane.

Articulation
Between a facet on the head of the fibula and a similar facet on the lateral condyle of the tibia.

Movements
Movement is slight and passively occurs along with movements of the ankle joint.

Distal Tibiofibular Joint

Type of Joint
Syndesmosis.

Articulation
Between the rough, triangular, opposed surfaces at the distal end of the tibia and fibula.

Movements
Movement is slight and passively occurs along with movements of the ankle joint.

a)

b)

Anterior
ligament
of fibula
head

Head of the
fibula

Interosseous
membrane

Tendon of
popliteus

Posterior ligament
of fibular head

Head of the fibula

Interosseous
membrane

Posterior
tibiofibular
ligament

Transverse
tibiofibular
ligament

Anterior and
posterior
tibiofibular
ligaments

c)

d)

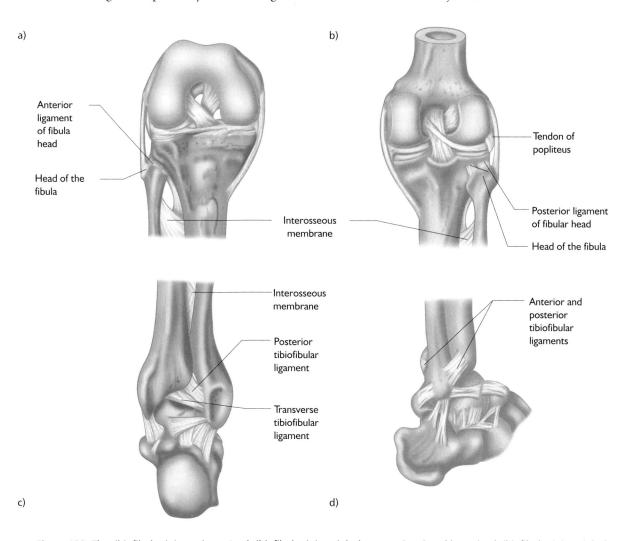

Figure 122: The tibiofibular joints; a) proximal tibiofibular joint, right leg, anterior view, b) proximal tibiofibular joint, right leg, posterior view, c) distal tibiofibular joint, right leg, posterior view, d) distal tibiofibular joint, right leg, lateral view.

Ankle Joint

Type of Joint
Synovial hinge.

Articulation
Between the distal tibia, the medial malleolus of the tibia, the lateral malleolus of the fibula and the talus. Therefore, the lower ends of the tibia and fibula provide a socket for the talus.

Movements
Dorsiflexion and plantar flexion.

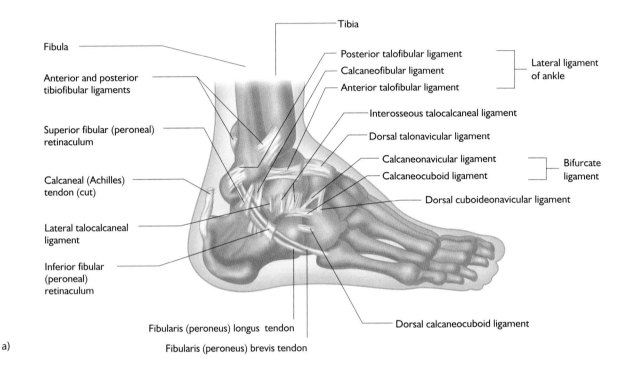

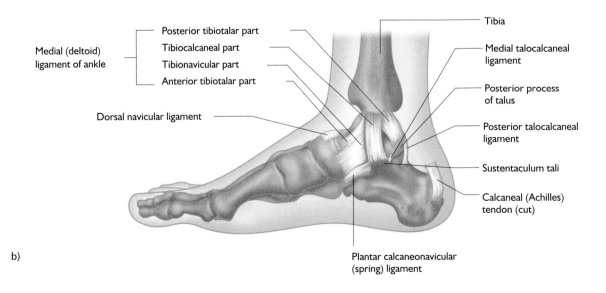

Figure 123: The ankle joint; a) right foot, lateral view, b) right foot, medial view.

The Arches of the Foot

The longitudinal arch:

- A series of synovial plane joints.
- It extends from the calcaneus to the metatarsals via the talus, navicular and cuneiforms.
- It is formed by the shapes of the metatarsal bones.
- It is supported by the calcaneonavicular (spring) ligament, a number of small interosseous ligaments, and the tendons of the tibialis anterior and tibialis posterior muscles.
- The arch is higher on the medial side than the lateral side.

The transverse arch:

- A series of synovial plane joints.
- It is placed through the distal row of tarsal bones.
- It is supported by the shape of the tarsal bones, many small interosseous ligaments and the tendons of the peroneus longus, tibialis anterior and tibialis posterior muscles.

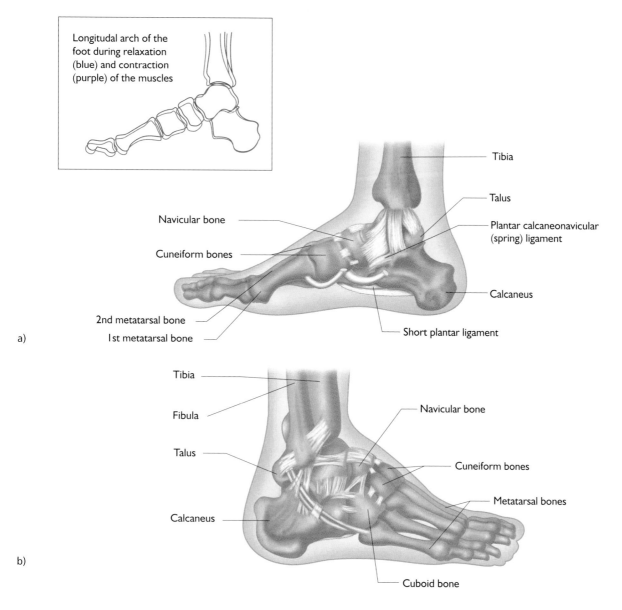

Figure 124: The arches of the foot; a) right foot, medial view, b) right foot, lateral view.

Intertarsal Joints

Type of Joints
A complex set of synovial plane joints.

Articulation
Subtalar joint: Between the inferior surface the talus and the superior surface of the calcaneus.
Talocalcaneonavicular joint: Between the talus, calcaneus and navicular.
Calcaneocuboid joint: Between the calcaneus and cuboid.
Transverse tarsal joint: A term to describe the transverse plane extending across the full width of the tarsus, comprising the talocalcaneonavicular joint and the calcaneocuboid joint.
Cuneonavicular joint: Between the cuneiform and the navicular.
Intercuneiform joints: Between the three cuneiform bones.
Cuneocuboid joint: Between the lateral cuneiform bone and the cuboid bone.

Movements of the Tarsus
Inversion and eversion of the foot.

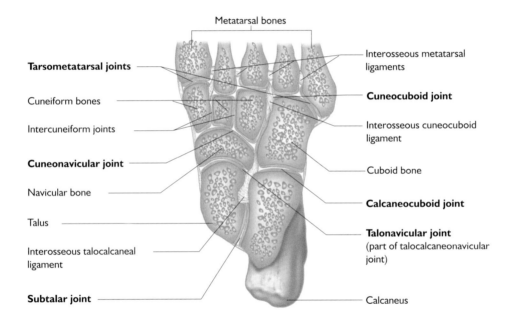

Figure 125: The intertarsal joints (horizontal section of right foot).

Tarsometatarsal and Intermetatarsal Joints

Type of Joints
Synovial plane.

Articulation
Tarsometatarsal joints: Between the distal (anterior) row of tarsal bones (the cuboid and three cuneiforms) and the bases of the metatarsal bones.
Intermetatarsal joints: Between facets on adjacent sides of the bases of all lateral metatarsal bones.

Movements
Small gliding movements of the metatarsals, limited by ligaments and the interlocking of the bones, contribute slightly to inversion and eversion of the foot.

Metatarsophalangeal Joints

Type of Joint
Synovial condyloid.

Articulation
Between the head of a metatarsal and the base of a proximal phalanx.

NOTE: The capsule is deficient on the dorsal aspect, where it is replaced by an expansion of the extensor tendon.

Movements
Flexion and extension. Abduction and adduction. Combined movements may produce passive circumduction.

NOTE: In flexion, the toes are drawn together; in extension they tend to spread apart and incline slightly laterally. Movements are less extensive than at the corresponding joints of the hand.

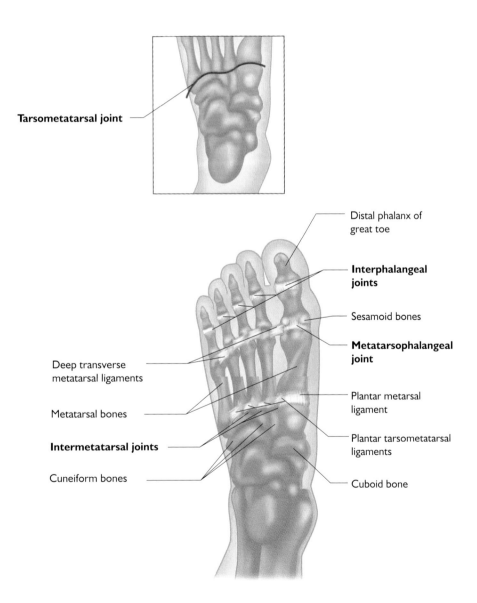

Figure 126: The tarsometatarsal, intermetatarsal and metatarsophalangeal joints (plantar view).

Interphalangeal Joints

Type of Joint
Synovial hinge.

Articulation
Between the proximal and middle phalanges (proximal interphalangeal joint, abbreviated PIP), or the middle and distal phalanges (distal interphalangeal joint, abbreviated DIP).

Movements
Flexion and extension.

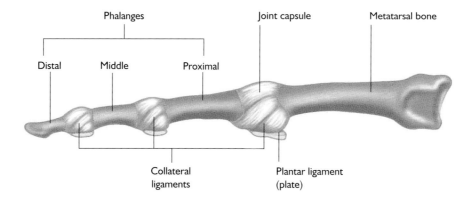

Figure 127: The metatarsophalangeal and interphalangeal joints (lateral view).

Skeletal Muscle and Fascia

7

Skeletal Muscle Structure and Function

Musculo-skeletal Mechanics

Skeletal Muscle Structure and Function

Skeletal (somatic or voluntary) muscles make up approximately 40% of the total human body weight. Their primary function is to produce movement through the ability to contract and relax in a coordinated manner. They are attached to bone by tendons. The place where a muscle attaches to a relatively stationary point on a bone, either directly or via a tendon, is called the *origin*. When the muscle contracts, it transmits tension to the bones across one or more joints, and movement occurs. The end of the muscle that attaches to the bone that moves is called the *insertion*.

Overview of Skeletal Muscle Structure

The functional unit of skeletal muscle is known as a *muscle fibre*, which is an elongated, cylindrical cell with multiple nuclei, ranging from 10 to 100 microns in width, and a few millimetres to 30+ centimetres in length. The cytoplasm of the fibre is called the *sarcoplasm*, which is encapsulated inside a cell membrane called the *sarcolemma*. A delicate membrane known as the *endomysium* surrounds each individual fibre.

These fibres are grouped together in bundles covered by the *perimysium*. These bundles are themselves grouped together, and the whole muscle is encased in a sheath called the *epimysium*. These muscle membranes lie through the entire length of the muscle, from the tendon of origin to the tendon of insertion. This whole structure is sometimes referred to as the *musculo-tendinous unit*.

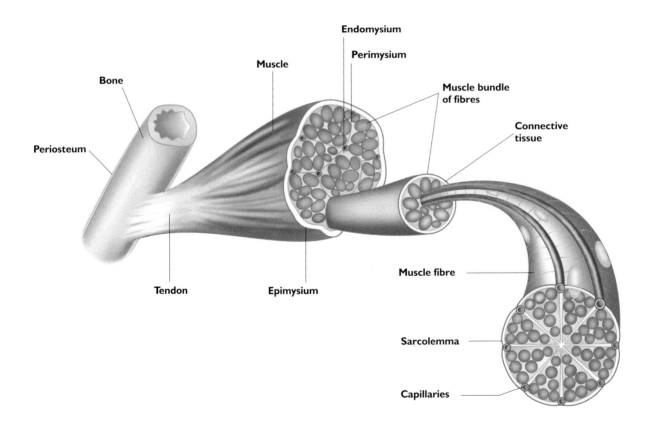

Figure 128: Cross-section of muscle tissue.

So, in defining the structure of muscle tissue in more detail, from the minute to gross, we have the following components:

Myofibrils

Through an electron microscope, one can distinguish the contractile elements of a muscle fibre, known as myofibrils, running the whole length of the fibre. Each myofibril reveals alternate light and dark banding, producing the characteristic cross-striation of the muscle fibre. These bands are called *myofilaments*. The light bands are referred to as isotropic (I) bands, and consist of thin myofilaments made of the protein actin. The dark one's are called anisotropic (A) bands, consisting of thicker myofilaments made of the protein myosin. (Note that a third connecting filament made of a protein called titin is now recognised). The myosin filaments have paddle-like extensions that emanate from the filaments rather like the oars of a boat. These extensions latch on to the actin filaments, forming what are described as 'cross-bridges' between the two types of filaments. The cross-bridges, using the energy of ATP, pull the actin strands closer together*. Thus, the light and dark sets of filaments increasingly overlap, like the interlocking of fingers, resulting in muscle contraction. One set of actin-myosin filaments is called a *sarcomere*.

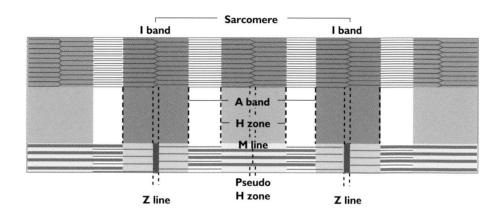

Figure 129: The myofilaments in a sarcomere. A sarcomere is bounded at both ends by the Z line.

- The lighter zone is known as the I band, and the darker zone the A band.
- The Z line is a thin dark line at the midpoint of the I band.
- A sarcomere is defined as the section of myofibril between one Z line and the next.
- The centre of the A band contains the H zone.
- The M line bisects the H zone, and delineates the centre of the sarcomere.

If an outside force causes a muscle to stretch beyond its resting level of tonus, the interlinking effect of the actin and myosin filaments that occurs during contraction is reversed. Initially, the actin and myosin filaments accommodate the stretch, but as the stretch continues, the titin filaments increasingly 'pay out' to absorb the displacement. Thus, it is the titin filament that determines the muscle fibre's extensibility and resistance to stretch. Research indicates that a muscle fibre (sarcomere), if properly prepared, can be elongated up to 150% of its normal length at rest.

Endomysium

A delicate connective tissue called endomysium lies outside the sarcolemma of each muscle fibre, separating each fibre from its neighbours, but also connecting them together.

---

* Huxley's Sliding Filament Theory
The generally accepted hypothesis to explain muscle function is partly described by Huxley's sliding filament theory (Huxley and Hanson, 1954). Muscle fibres receive a nerve impulse that cause the release of calcium ions stored in the muscle. In the presence of the muscles fuel, known as adenosine triphosphate (ATP), the calcium ions bind with the actin and myosin filaments to form an electrostatic (magnetic) bond. This bond causes the fibres to shorten, resulting in their contraction or increase in tonus. When the nerve impulse ceases, the muscle fibres relax. Their elastic elements recoil the filaments to their non-contracted lengths, i.e. their resting level of tonus.

Fasciculi
Muscle fibres are arranged in parallel bundles called fasciculi.

Perimysium
Each fasciculus is bound by a denser collagenic sheath called the perimysium.

Epimysium
The entire muscle, which is therefore an assembly of fasciculi, is wrapped in a fibrous sheath called the epimysium.

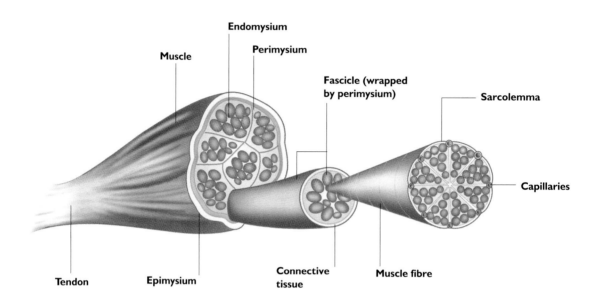

Figure 130: The connective tissue sheaths of skeletal muscle.

Deep Fascia
A coarser sheet of fibrous connective tissue lies outside the epimysium, binding individual muscles into functional groups. This deep fascia extends to wrap around other adjacent structures.

Muscle Attachment
The way a muscle attaches to bone or other tissues is either through a direct attachment or an indirect attachment. A direct attachment (called a fleshy attachment) is where the perimysium and epimysium of the muscle unite and fuse with the periosteum of a bone, perichondrium of a cartilage, a joint capsule, or the connective tissue underlying the skin (some muscles of facial expression being good examples of the latter). An indirect attachment is where the connective tissue components of a muscle fuse together into bundles of collagen fibres to form an intervening tendon. Indirect attachments are much more common. The types of tendinous attachments are as follows:

Tendons and Aponeurosis
Muscle fascia, which is the connective tissue component of a muscle, combine together and extend beyond the end of the muscle as round cords or flat bands, called tendons; or as a thin, flat and broad aponeurosis. The tendon or aponeurosis secures the muscle to the bone or cartilage, to the fascia of other muscles, or to a seam of fibrous tissue called a *raphe*. Flat patches of tendon may form on the body of a muscle where it is exposed to friction. For example, on the deep surface of trapezius, where it rubs against the spine of the scapula.

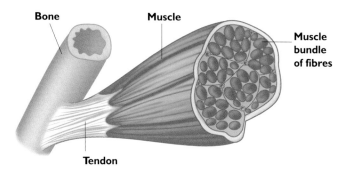

Figure 131: A tendon attachment.

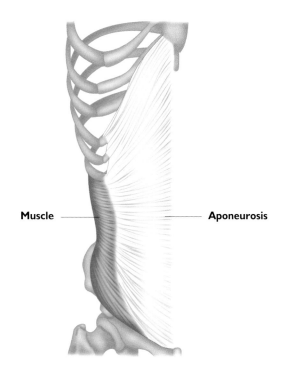

Figure 132: An attachment by aponeurosis.

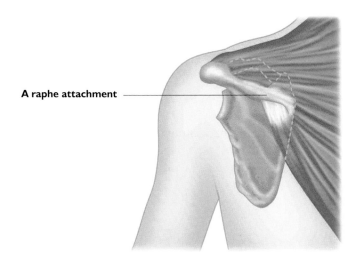

Figure 133: Flat patches of tendon on the deep surface of trapezius.

Intermuscular Septa

In some cases, flat sheets of dense connective tissue known as intermuscular septa penetrate between muscles, providing another medium to which muscle fibres may attach.

Sesamoid Bones

If a tendon is subject to friction, it may, but not necessarily, develop a sesamoid bone within its substance. An example is the peroneus longus tendon in the sole of the foot. However, sesamoid bones may also appear in tendons not subject to friction.

Multiple Attachments

Many muscles have only two attachments, one at each end. However, more complex muscles are often attached to several different structures at its origin and/or its insertion. If these attachments are separated, effectively meaning the muscle gives rise to two or more tendons and/or aponeurosis inserting into different places, the muscle is said to have two heads. For example, the biceps brachii has two heads at its origin; one from the corocoid process of the scapula and one the other from the supraglenoid tubercle (*see* p.272). The triceps has three heads and the quadriceps has four.

Red and White Muscle Fibres

There are three types of skeletal muscle fibres: red slow-twitch fibres, white fast-twitch fibres, and intermediate fast twitch fibres.

1. Red Slow-twitch Fibres

These are thin cells that contract slowly. The red colour is due to their content of myoglobin, a substance similar to haemoglobin, which stores oxygen and increases the rate of oxygen diffusion within the muscle fibre. As long as oxygen supply is plentiful, red fibres can contract for sustained periods, and are thus very resistant to fatigue. Successful marathon runners tend to have a high percentage of these red fibres.

2. White Fast-twitch Fibres

These are large cells that contract rapidly. They are pale, due to a lesser content of myoglobin. They fatigue quickly, because they rely on short-lived glycogen reserves in the fibre to contract. However, they are capable of generating much more powerful contractions than red fibres, enabling them to perform rapid, powerful movements for short periods. Successful sprinters have a higher proportion of these white fibres.

3. Intermediate Fast-twitch Fibres

These red or pink fibres are a compromise in size and activity between the red and white fibres.

NOTE: There is always a mixture of these muscle fibres in any given muscle, giving them a range of fatigue resistance and contractile speeds

Blood Supply

In general, each muscle receives one artery to *bring* nutrients via blood into the muscle, and several veins, to *take away* metabolic waste products surrendered by the muscle into the blood. These blood vessels generally enter through the central part of the muscle, but can also enter towards one end. Thereafter, they branch into a capillary plexus, which spreads throughout the intermuscular septa, to eventually penetrate the endomysium around each muscle fibre. During exercise the capillaries dilate, increasing the amount of blood flow in the muscle by up to 800 times. The muscle tendon, because it is composed of a relatively inactive tissue, has a much less extensive blood supply.

Nerve Supply

The nerve supply to a muscle usually enters at the same place as the blood supply, and branches through the connective tissue septa into the endomysium in a similar way. Each skeletal muscle fibre is supplied by a single nerve ending. This is in contrast to other muscle tissues, which are able to contract without any nerve stimulation.

The nerve entering the muscle usually contains roughly equal proportions of sensory and motor nerve fibres, although some muscles may receive separate sensory branches. As the nerve fibre approaches the muscle fibre, it divides into a number of terminal branches, collectively called a *motor end plate*.

Motor Unit of a Skeletal Muscle

A motor unit consists of a single motor nerve cell and the muscle fibres stimulated by it. The motor units vary in size, ranging from cylinders of muscle 5–7mm in diameter in the upper limb and 7–10mm in diameter in the lower limb. The average number of muscle fibres within a unit is 150 (but this number ranges from less than 10 to several hundred). Where fine gradations of movement are required, as in the muscles of the eyeball or fingers, the number of muscle fibres supplied by a single nerve cell is small. On the other hand, where more gross movements are required, as in the muscles of the lower limb, each nerve cell may supply a motor unit of several hundred fibres.

The muscle fibres in a single motor unit are spread throughout the muscle, rather than being clustered together. This means that stimulation of a single motor unit will cause the entire muscle to exhibit a weak contraction.

Skeletal muscles work on an '*All or Nothing Principle*'. In other words, groups of muscle cells, or fasciculi, can either contract or not contract. Depending on the strength of contraction required, a certain number of muscle cells will contract totally, while others will not contract at all. When a great muscular effort is needed, most of the motor units may be stimulated at the same time. However, under normal conditions, the motor units tend to work in relays, so that during prolonged contractions some are resting while others are contracting.

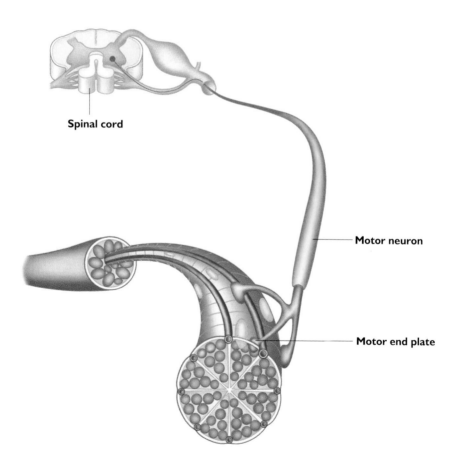

Spinal cord

Motor neuron

Motor end plate

Figure 134: A motor unit of a skeletal muscle.

Muscle Reflexes

Within skeletal muscles there are two specialized types of nerve receptors that can sense stretch. These are muscle spindles and Golgi tendon organs (GTO's). Muscle spindles are cigar-like in shape and consist of tiny modified muscle fibres called intrafusal fibres, and nerve endings, encased together within a connective tissue sheath. They lie between and parallel to the main muscle fibres. The GTO's are located mostly at the junction of muscles and their tendons or aponeurosis.

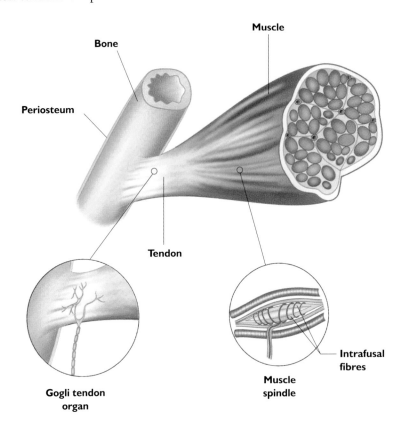

Figure 135: Anatomy of the muscle spindle and Golgi tendon organ.

Stretch Reflex

The stretch reflex helps control posture by maintaining muscle tone. It also helps prevent injury, by enabling a muscle to respond to a sudden or unexpected increase in length. The way it works is as follows:

1. When a muscle is lengthened, the muscle spindles are also stretched, causing each spindle to send a nerve impulse to the spinal cord.

2. On receiving this impulse, the spinal cord immediately sends an impulse back to the stretched muscle fibres, causing them to contract, in order to resist further stretching of the muscle. This circular process is known as a *reflex arc*.

3. An impulse is simultaneously sent from the spinal cord to the antagonist of the contracting muscle (i.e. the muscle opposing the contraction), causing the antagonist to relax; so that it cannot resist the contraction of the stretched muscle. This process is known as *reciprocal inhibition*.

4. Concurrent with this spinal reflex, nerve impulses are also sent up the spinal cord to the brain to relay information on muscle length and the speed of muscle contraction. A reflex in the brain feeds nerve impulses back to the muscle to ensure the appropriate muscle tone is maintained to meet the requirements of posture and movement.

5. Meanwhile, the stretch sensitivity of the minute intrafusal muscle fibres within the muscle spindle are smoothed and regulated by *gamma efferent nerve fibres*, arising from motor neurons within the spinal cord. Thus, a *gamma motor neuron reflex arc* ensures the evenness of muscle contraction, which would otherwise be jerky if muscle tone relied on the stretch reflex alone.

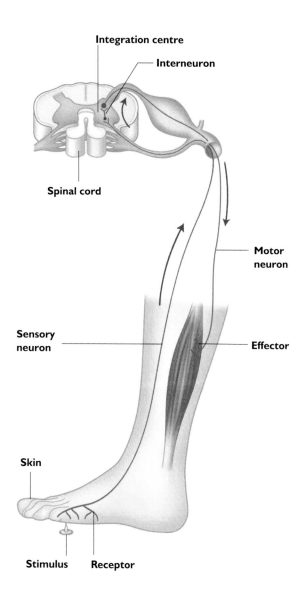

Figure 136: The basic reflex arc.

The classic clinical example of the stretch reflex in action is the knee jerk, or patellar reflex; whereby the patellar tendon is lightly struck with a small rubber hammer. This results in the following sequence of events:

1. The sudden stretch of the patellar tendon causes the quadriceps to be stretched.

2. This rapid stretch is registered by the muscle spindles within the quadriceps, causing the quadriceps to contract. This causes a small kick as the knee straightens suddenly, and takes the tension off the muscle spindles.

3. Simultaneously, nerve impulses to the hamstrings, which are the antagonists of the quadriceps, are inhibited, causing the hamstrings to relax.

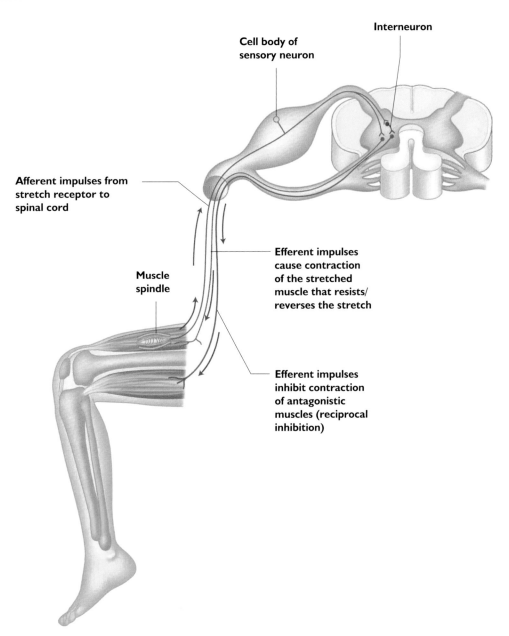

Interneuron

Cell body of sensory neuron

Afferent impulses from stretch receptor to spinal cord

Efferent impulses cause contraction of the stretched muscle that resists/ reverses the stretch

Muscle spindle

Efferent impulses inhibit contraction of antagonistic muscles (reciprocal inhibition)

Figure 137: The stretch reflex arc.

Another obvious example of the stretch reflex in action is: when a person falls asleep in the sitting position, the head will relax forward, then jerk back up, because the stretched muscle spindles in the back of the neck have activated a reflex arc.

The stretch reflex also works constantly to maintain the tonus of our postural muscles. That is, it enables us to remain standing without conscious effort and without collapsing forwards. The sequence of events preventing this forward collapse occurs in a fraction of a second, as follows:

1. In standing, we naturally begin to sway forwards.

2. This pulls our calf muscles into a lengthened position, activating the stretch reflex.

3. The calf muscles consequently contract to pull us back to the upright position.

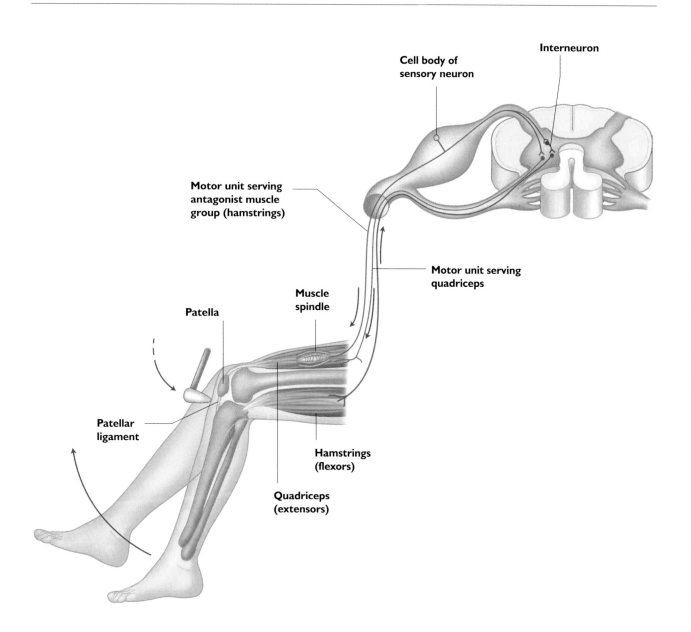

Figure 138: The patellar reflex.

Deep Tendon Reflex (Autogenic Inhibition)

In contrast to the stretch reflex, which involves the muscle spindle's response to muscle elongation, the deep tendon reflex involves the reaction of Golgi tendon organs (GTO's) to muscle contraction. As such, the deep tendon reflex creates the opposite effect to the stretch reflex. The way it works is as follows:

1. When a muscle contracts, it pulls on the tendons which are situated at either end of the muscle.

2. The tension in the tendon causes the GTO's to transmit impulses to the spinal cord, (some impulses continue on to the cerebellum).

3. As these impulses reach the spinal cord, they inhibit the motor nerves supplying the contracting muscle, causing it to relax.

4. Simultaneously, the motor nerves supplying the antagonist muscle are activated, causing it to contract. This process is called *reciprocal activation*.

5. Meanwhile, the information reaching the cerebellum is processed and fed back to help readjust muscle tension.

The deep tendon reflex has a protective function, preventing the muscle from contracting so hard that it rips its attachment off the bone. It is therefore especially important during activities such as running, which involve rapid switching between flexion and extension.

Note however, that in normal day-to-day movement, tension in the muscles is not sufficient to activate the GTO's deep tendon reflex. By contrast, the threshold of the muscle spindle stretch reflex is set much lower, because it must constantly maintain sufficient tonus in the postural muscles to keep us upright. Hence, it is active throughout normal daily activity.

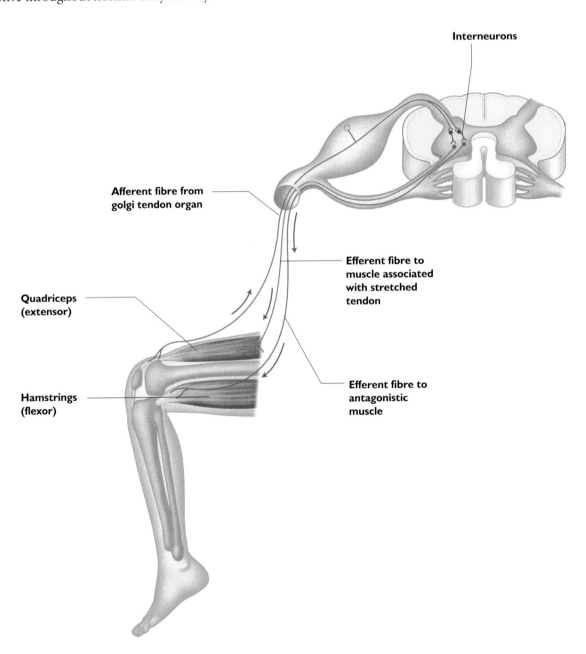

Figure 139: The deep tendon reflex.

Isometric and Isotonic Contractions

A muscle will contract upon stimulation, in an attempt to bring its attachments closer together, but this does not necessarily result in a shortening of the muscle. If the contraction of muscle results in the muscle creating movement of some sort, the contraction is called isotonic. If no movement results from contraction, such a contraction is called isometric.

Isometric

An isometric contraction occurs when a muscle increases its tension, but the length of the muscle is not altered. In other words, although the muscle tenses, the joint over which the muscle works does not move. One example of this is holding a heavy object in the hand with the elbow held stationary and bent at 90 degrees. Trying to lift something that proves to be too heavy to move is another example. Note also that some of the postural muscles are largely working isometrically by automatic reflex. For example, in the upright position, the body has a natural tendency to fall forward at the ankle. This is prevented by isometric contraction of the calf muscles. Likewise, the centre of gravity of the skull would make the head tilt forwards if the muscles at the back of the neck did not contract isometrically to keep the head centralized.

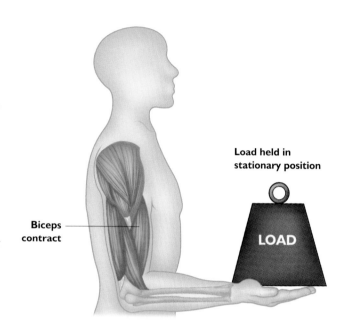

Figure 140: Isometric contraction.

Isotonic

It is the isotonic contractions of muscle that enable us to move about. Such contractions are of two types:

Eccentric

Eccentric contraction means that the muscle fibres 'pay out' in a controlled manner to slow down movements which gravity, if unchecked, would otherwise cause to be too rapid. For example, lowering an object held in the hand down to your side. Another example is simply sitting down into a chair. Therefore, the difference between concentric and eccentric contraction is that in the former, the muscle shortens, and in the latter, it actually lengthens.

Concentric

In concentric contractions, the muscle attachments move closer together, causing movement at the joint. Using the example of holding an object in the hand, if the biceps muscle contracts concentrically, the elbow joint will flex and the hand will move towards the shoulder. Similarly, if we look up at the stars, the muscles at the back of the neck must contract concentrically to tilt the head back and extend the neck.

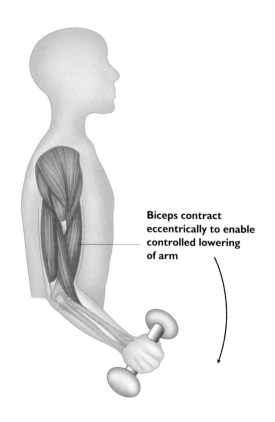

Figure 141: Eccentric isotonic contraction.

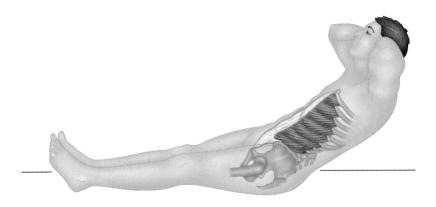

Figure 142: Abdominals contract to raise body concentrically.

Muscle Shape (Arrangement of Fascicles)

Muscles come in a variety of shapes according to the arrangement of their fascicles. The reason for this variation is to provide optimum mechanical efficiency for a muscle in relation to its position and action. The most common arrangement of fascicles give muscle shapes described as parallel, pennate, convergent and circular. Each of these shapes has further sub-categories.

Parallel

This arrangement has the fascicles running parallel to the long axis of the muscle. If the fascicles extend throughout the length of the muscle, it is known as a *strap muscle*, for example: sartorius (*see* p.333) or sternohyoideus (*see* p.189). If the muscle also has an expanded belly and tendons at both ends, it is called a *fusiform* muscle, for example, the biceps brachii of the arm (*see* p.272). A modification of this type of muscle has a fleshy belly at either end, with a tendon in the middle. Such muscles are referred to as *digastric*, e.g. digastricus (*see* p.188).

Pennate

Pennate muscles are so named because their short fasciculi are attached obliquely to the tendon, like the structure of a feather (penna = feather). If the tendon develops on one side of the muscle, it is referred to as *unipennate*, for example, the flexor digitorum longus in the leg (*see* p.351). If the tendon is in the middle and fibres are attached obliquely from both sides, it is known as *bipennate*, of which the rectus femoris is a good example (*see* p.334). If there are numerous tendinous intrusions into the muscle with fibres attaching obliquely from several directions, thus resembling many feathers side by side, the muscle is referred to as *multipennate*; the best example being the middle part of the deltoid muscle (*see* p.265).

Convergent

Muscles that have a broad origin with fascicles converging toward a single tendon, giving the muscle a triangular shape, are called convergent muscles. The best example is the pectoralis major (*see* p.262).

Circular

When the fascicles of a muscle are arranged in concentric rings, the muscle is referred to as *circular*. All the sphincter skeletal muscles in the body are of this type; i.e. they surround openings, which they close by contracting. Examples include the orbicularis oculi (*see* p.159) and orbicularis oris (*see* p.167).

When a muscle contracts, it can shorten by up to 70% of its original length. Hence, the longer the fibres, the greater the range of movement. On the other hand, the strength of a muscle depends on the total number of muscle fibres it contains, rather than their length. Therefore:

1. Muscles with long parallel fibres produce the greatest range of movement, but are not usually very powerful.
2. Muscles with a pennate pattern, especially if multipennate, pack in the most fibres. Such muscles shorten less than long parallel muscles, but tend to be much more powerful.

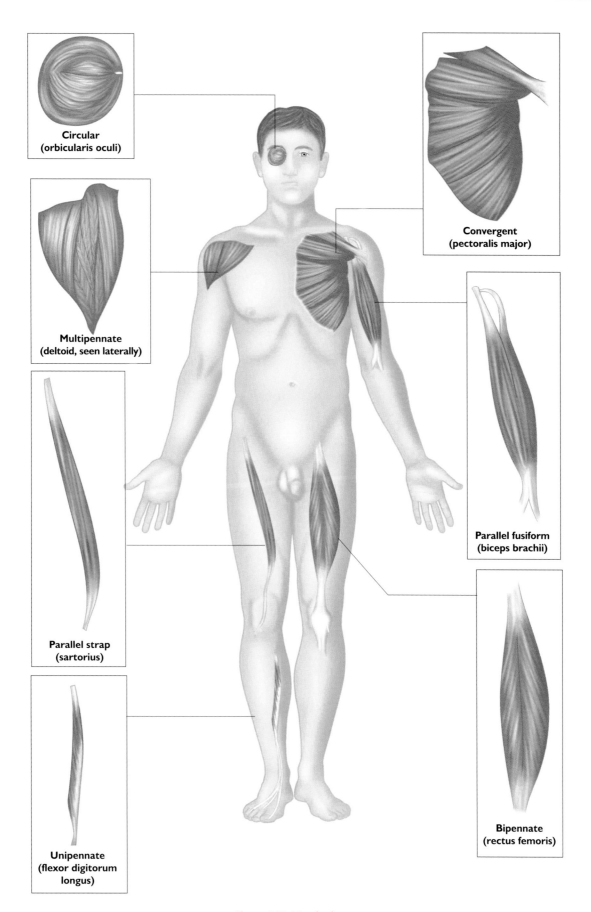

Circular
(orbicularis oculi)

Convergent
(pectoralis major)

Multipennate
(deltoid, seen laterally)

Parallel fusiform
(biceps brachii)

Parallel strap
(sartorius)

Bipennate
(rectus femoris)

Unipennate
(flexor digitorum
longus)

Figure 143: Muscle shapes.

Functional Characteristics of a Skeletal Muscle

All that has been said about muscles so far in this book enables us to formulate a list of functional characteristics pertaining to skeletal muscle.

Excitability

Excitability is the ability to receive and to respond to a stimulus. In the case of a muscle, when a nervous impulse from the brain reaches the muscle, a chemical known as *acetylcholine* is released. This chemical produces a change in the electrical balance in the muscle fibre and as a result generates an electrical current, known as an *action potential*. The action potential conducts the electrical current from one end of the muscle cell to the other and results in a contraction of the muscle cell, or muscle fibre (remember that one muscle cell = one muscle fibre).

Contractility

Contractility is the ability of a muscle to shorten forcibly when stimulated. In other words, the muscles themselves can only contract. They cannot lengthen, except via some external means (i.e. manually), beyond their normal resting length (*see Tonus* below). In other words, muscles can only pull their ends together (contract); they cannot push them apart.

Extensibility

Extensibility is the ability of a muscle to be extended, or returned to its resting length (which is a semi-contracted state), or slightly beyond. For example, if we bend forward at the hips from standing, the muscles of the back, such as erector spinae, lengthen eccentrically (*see p.131*) to lower the trunk, paying out slightly beyond their normal resting length, and are thus effectively 'elongated'.

Elasticity

Elasticity describes the ability of a muscle fibre to recoil after being lengthened, and therefore resume its resting length when relaxed. In a whole muscle, the elastic effect is supplemented by the important elastic properties of the connective tissue sheaths (endomysium and epimysium). Tendons also contribute some elastic properties. An example of this elastic recoil effect can be experienced when coming back up from a forward bend at the hips as described above, there is initially no muscle contraction. Instead, the upward movement is initiated purely by elastic recoil of the back muscles, after which, contraction of the back muscles completes the movement.

Tonus

Tonus, or muscle tone, is the term used to describe the slightly contracted state which muscles resume during the resting state. Muscle tonus does not produce active movements, but it keeps the muscles firm, healthy, and ready to respond to stimulation. It is the tonus of skeletal muscles that also helps stabilize and maintain posture. *Hypertonic* muscles are those muscles whose 'normal' resting state is over-contracted.

General Functions of Skeletal Muscles

Enable Movement

Skeletal muscles are responsible for all locomotion and manipulation, and they enable you to respond quickly.

Maintain Posture

Skeletal muscles support an upright posture against the pull of gravity.

Stabilize Joints

Skeletal muscles and their tendons stabilize joints.

Generate Heat

In common with smooth and cardiac muscles, skeletal muscles generate heat, which is important in maintaining a normal body temperature.

Musculo-skeletal Mechanics

Origins and Insertions

In the majority of movements, one attachment of a muscle remains relatively stationary while the attachment at the other end moves. The more stationary attachment is called the *origin* of the muscle, and the other attachment is called the *insertion*. A spring that closes a gate could be said to have its origin on the gate-post and its insertion on the gate itself. In the body, the arrangement is rarely so clear-cut, because depending on the activity one is engaged in, the fixed and moveable ends of the muscle may be reversed. For example, muscles that attach the upper limb to the chest normally move the arm relative to the trunk; which means their origins are on the trunk and their insertions are on the upper limb. However, in climbing, the arms are fixed, while the trunk is moved as it is pulled up to the fixed limbs. In this type of situation, where the insertion is fixed and the origin moves, the muscle is said to perform a *reversed action*. Because there are so many situations where muscles are working with a reversed action, it is sometimes less confusing to simply speak of 'attachments', without reference to origin and insertion.

In practice, muscle attachments that lie more proximally, i.e. more towards the trunk or on the trunk, are usually referred to as the origin. Attachments that lie more distally, i.e. away from the attached end of a limb, or away from the trunk, are referred to as the insertion.

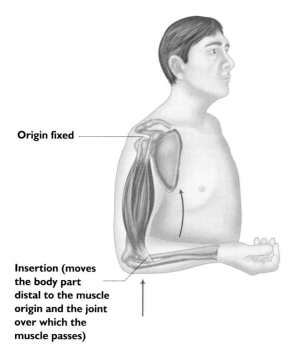

Origin fixed

Insertion (moves the body part distal to the muscle origin and the joint over which the muscle passes)

Figure 144: Muscle working with origin fixed and insertion moving.

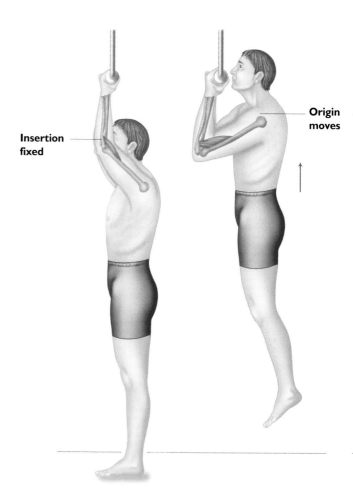

Insertion fixed

Origin moves

Figure 145: Climbing: muscles are working with insertion fixed and origin moving (reversed action).

Group Action of Muscles

Muscles work together, or in opposition, to achieve a wide variety of movements. Therefore, whatever one muscle can do, there is another muscle that can undo it. Muscles may also be required to provide additional support or stability to enable certain movements to occur elsewhere.

Muscles are classified into four functional groups:

• Prime mover or Agonist
• Antagonist
• Synergist
• Fixator

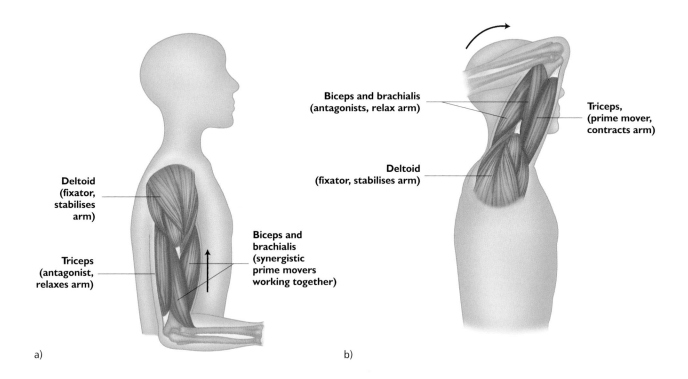

Figure 146: Group action of muscles; a) flexing arm at elbow,
b) extending arm at elbow (showing reversed roles of prime mover and antagonist).

Prime Mover or Agonist

A prime mover (also called an agonist) is a muscle that contracts to produce a specified movement. An example is the biceps brachii, which is the prime mover of elbow flexion. Other muscles may assist the prime mover in providing the same movement, albeit with less effect. Such muscles are called assistant or secondary movers. For example, the brachialis assists the biceps brachii in flexing the elbow, and is therefore a secondary mover.

Antagonist

The muscle on the opposite side of a joint to the prime mover, and which must relax to allow the prime mover to contract, is called an antagonist. For example, when the biceps on the front of the arm contract to flex the elbow, the triceps on the back of the arm must relax to allow this movement to occur. When the movement is reversed, i.e. when the elbow is extended, the triceps becomes the prime mover and the biceps assumes the role of antagonist.

Synergist

Synergists prevent any unwanted movements that might occur as the prime mover contracts. This is especially important where a prime mover crosses two joints, because when it contracts it will cause movement at both joints, unless other muscles act to stabilize one of the joints. For example, the muscles that flex the fingers not only cross the finger joints, but also cross the wrist joint, potentially causing movement at both joints. However, it is because you have other muscles acting synergistically to stabilize the wrist joint that you are able to flex the fingers into a fist without also flexing the wrist at the same time.

A prime mover may have more than one action, so synergists also act to eliminate the unwanted movements. For example, the biceps brachii will flex the elbow, but its line of pull will also supinate the forearm (twist the forearm, as in tightening a screw). If you want flexion to occur without supination, other muscles must contract to prevent this supination. In this context, such synergists are sometimes called neutralisers.

Fixator

A synergist is more specifically referred to as a fixator or stabilizer when it immobilizes the bone of the prime mover's origin, thus providing a stable base for the action of the prime mover. The muscles that stabilize (fix) the scapula during movements of the upper limb are good examples. The sit-up exercise gives another good example: The abdominal muscles attach to both the ribcage and the pelvis. When they contract to enable you to perform a sit-up, the hip flexors will contract synergistically as fixators to prevent the abdominals tilting the pelvis; enabling the upper body to curl forward as the pelvis remains stationary.

Leverage

The bones, joints, and muscles together form a system of levers in the body, in order to optimise the relative strength, range and speed required of any given movement. The joints act as the fulcra (sing. fulcrum), while the muscles apply the effort and the bones bear the weight of the body part to be moved.

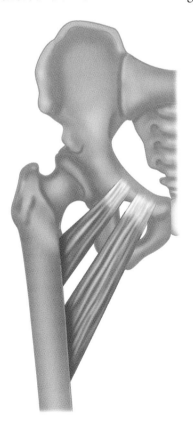

A muscle attached close to the fulcrum will be relatively weaker than it would be if it were attached further away. However, it is able to produce a greater range and speed of movement; because the length of the lever amplifies the distance travelled by its moveable attachment. Figure 147 illustrates this in relation to the adductors of the hip joint. The muscle so positioned to move the greater load (in this case, adductor longus) is said to have a *mechanical advantage*. The muscle attached close to the fulcra is said to operate at a *mechanical disadvantage*, although it can move a load more rapidly through larger distances.

Figure 147: The pectineus is attached closer to the axis of movement than the adductor longus. Therefore, the pectineus is the weaker adductor of the hip, but is able to produce a greater movement of the lower limb per centimetre of contraction.

The following illustrations depict the differences in first, second and third class levers, with examples in the human body.

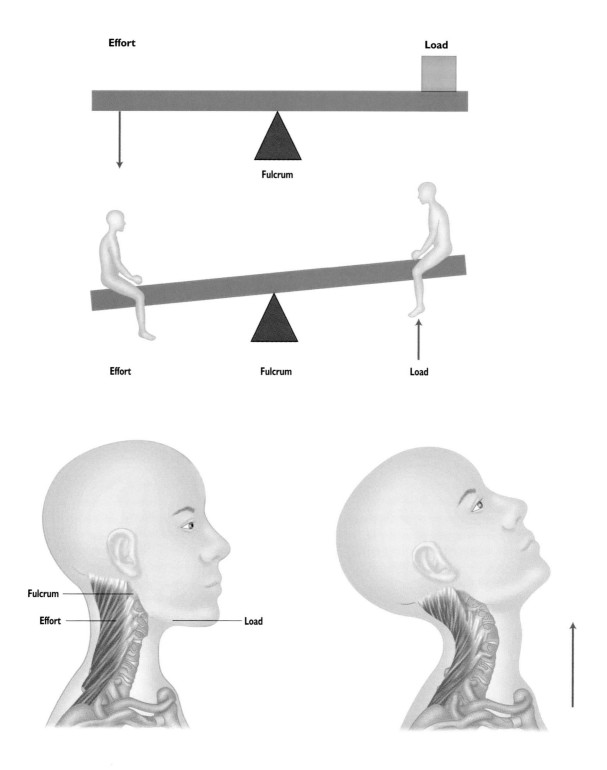

Fig 148: First-class lever: The relative position of components is Load-Fulcrum-Effort. Examples include a seesaw (as above). Another example is a pair of scissors. In the body, an example is the ability to extend the head and neck, i.e. the facial structures are the load; the atlanto-occipital joint is the fulcrum; the posterior neck muscles provide the effort.

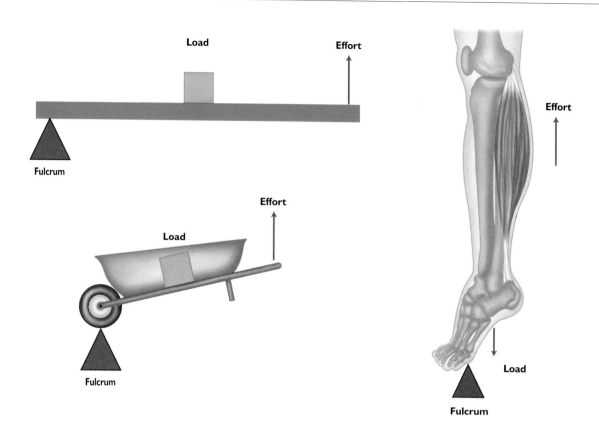

Figure 149: Second-class lever: The relative position of components is Fulcrum-Load-Effort. The best example is a wheelbarrow. In the body, an example is the ability to raise the heels off the ground in standing, i.e. the ball of the foot is the fulcrum; the body-weight is the load; the calf muscles provide the effort. With second-class levers, speed and range of movement is sacrificed for strength.

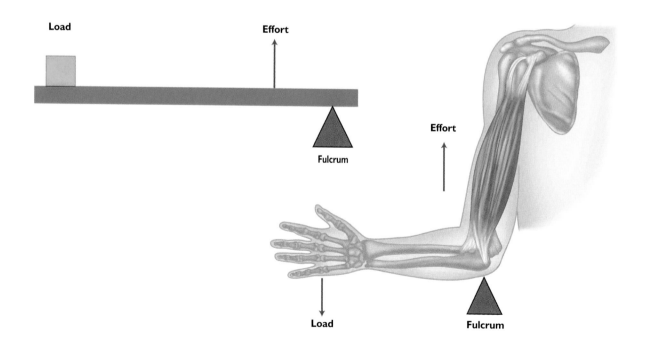

Figure 150: Third-class lever: The relative position of components is Load-Effort-Fulcrum. A pair of tweezers is an example of this. In the body, most skeletal muscles act in this way. An example is flexing the forearm, i.e. an object held in the hand is the load; the biceps provide the effort; the elbow joint is the fulcrum. With third-class levers, strength is sacrificed for speed and range of movement.

Factors in Muscles that Limit Skeletal Movement

The inability of a muscle to contract or lengthen beyond a certain point can cause some practical hindrances to bodily movement; which are outlined as follows:

Passive Insufficiency

Muscles that span over two joints are called biarticular muscles. These muscles may be unable to 'pay out' sufficiently to allow full movement of both joints simultaneously, unless the muscle has been trained to relax. For example, most people need to bend their knees in order to touch their toes. This is because the hamstrings (which span the hip and knee joints) cannot lengthen enough to allow full flexion at the hip joint without also pulling the knee joint into flexion. For the same reason, it is easier to pull your thigh to your chest if your knee is bent than it is with your knee straight. This limitation is called *passive insufficiency*. Passive insufficiency is therefore the inability of a muscle to lengthen by more than a fixed percentage of its length.

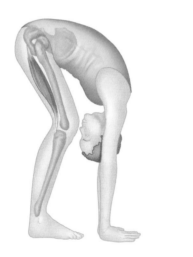
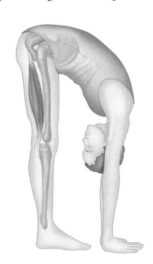

a)   b)

Fig 151: Passive insufficiency example 1; a) having to bend the knees to the touch toes means there is passive insufficiency of the hamstrings, and b) being able to touch the toes with knees straight means there is much less passive insufficiency of the hamstrings.

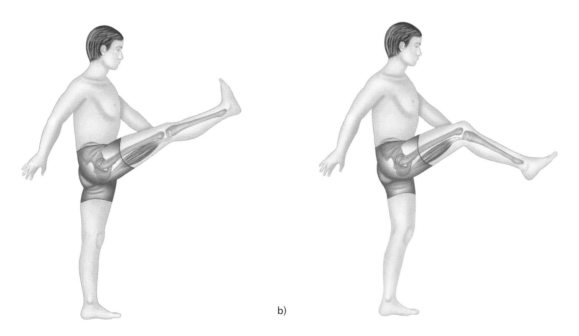

a)   b)

Fig 152: Passive insufficiency example 2; a) a high kick with knee straight is possible only if the hamstrings have been trained to overcome their passive insufficiency, and b) for most people, an attempt at a high kick will be restricted by hamstring passive insufficiency causing the knee to bend.

Active Insufficiency

Active insufficiency is the opposite of passive insufficiency. Whereas passive insufficiency results from the inability of a muscle to *lengthen* by more than a fixed percentage of its length, active insufficiency results from the inability of a muscle to *contract* by more than a fixed amount. For example, most people can flex their knee to bring their heel close to their buttock, if their hip is flexed; because the upper part of the hamstrings are lengthened and the lower part is shortened. However, one is normally unable to fully flex the knee when the hip is extended. This is because with hip extended, the hamstrings are already shortened, meaning that there is insufficient 'shortening' potential remaining in the hamstrings to then fully flex the knee.

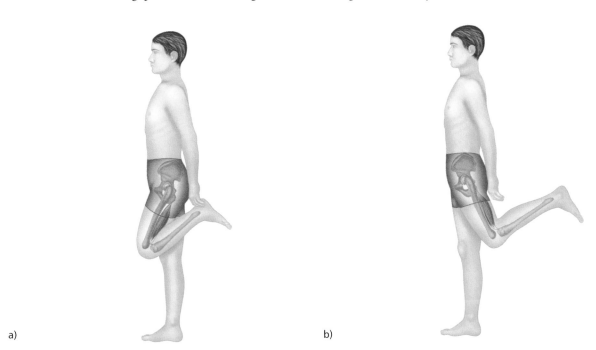

a)   b)

Fig 153: Active insufficiency; a) with hip flexed, the hamstrings are stretched at the hip, enabling its contraction to fully flex the knee, and b) with hip extended, the shortened hamstrings are unable to contract still further to fully flex the knee.

Concurrent Movement

If extension of the hip is required at the same time as extension of the knee, as in the push off from the ground in running, the phenomenon known as *concurrent movement* applies, and proves very useful. To grasp the concept of concurrent movement, first remember that when the hamstrings contract, they are able to both extend the hip joint and flex the knee joint, either singly or simultaneously. So, in analysing the example of running in more detail, we observe the following:

- As the foot pushes against the ground, the hamstrings contract to extend the hip.
- Meanwhile, fixators prevent the hamstrings from flexing the knee.
- Therefore, the hamstrings are shortened only at their upper end (origin), but remain lengthened at their lower end (insertion).
- The antagonist to the hamstring's action of flexing the hip is the rectus femoris, which relaxes because of reciprocal inhibition (*see* p.126) to allow the hamstrings to contract.
- When the hip is well extended, the already stretched rectus femoris is unable to lengthen further, causing it to pull the knee into extension.
- Therefore, the rectus femoris is lengthened at its upper end and shortened at its lower end.

Concurrent movement therefore avoids passive and active insufficiency of the hamstrings and rectus femoris by neither shortening nor stretching both ends of either muscle, but rather, having one end lengthen as the other shortens, and vice versa in the other muscle. Figure 154 should hopefully elucidate this concept.

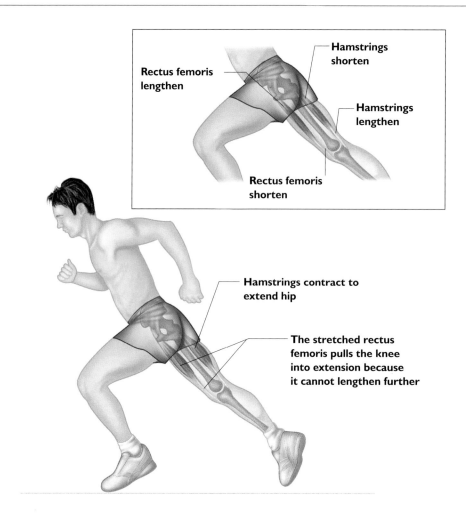

Figure 154: Concurrent movement.

Countercurrent Movement

If flexion of the hip is required to occur at the same time as extending the knee, as in kicking a ball, a countercurrent movement occurs. So, in analysing the example of kicking in more detail, we observe the following:

- To kick a ball, the rectus femoris acts as a prime mover to flex the hip and extend the knee.
- Thus, both the upper and lower portions of rectus femoris are shortened.
- The hamstrings relax due to reciprocal inhibition, so that they can extend at both ends and allow the kick to occur.
- The rectus femoris relaxes once the movement has been made, but the momentum of the movement is still propelling the leg forward.
- At this stage, the hamstrings contract to act as a 'brake' for the leg, as it flies forward.

Countercurrent movements therefore prevent injury by ensuring the antagonist relaxes first, then contracts at the right time to prevent the forces of momentum from overstretching muscles and ligaments. So called ballistic movements are relying on this principle, but are often done so forcefully that the power of momentum is greater than the ability of the antagonist to 'brake' that momentum. In such instances, muscle and ligament damage often occurs.

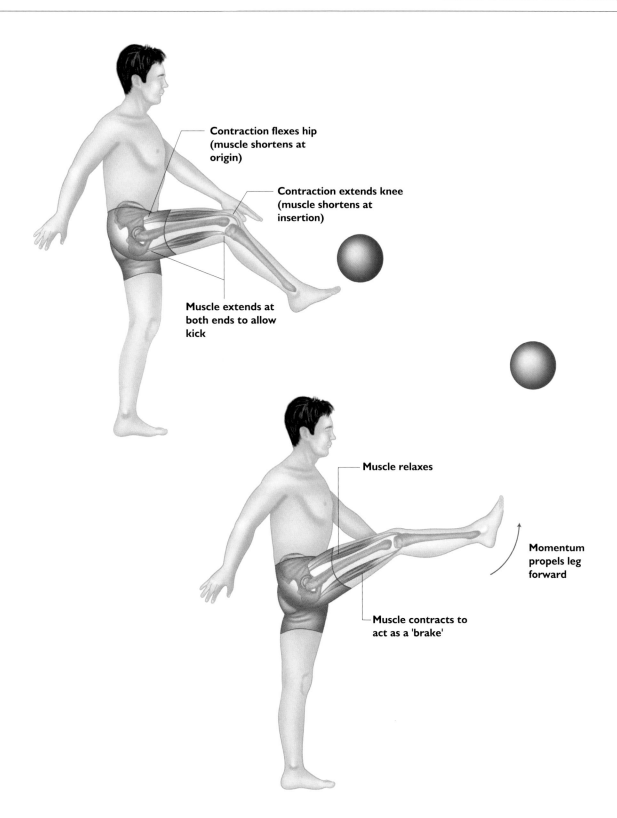

Contraction flexes hip
(muscle shortens at
origin)

Contraction extends knee
(muscle shortens at
insertion)

Muscle extends at
both ends to allow
kick

Muscle relaxes

Momentum
propels leg
forward

Muscle contracts to
act as a 'brake'

Figure 155: Countercurrent movement.

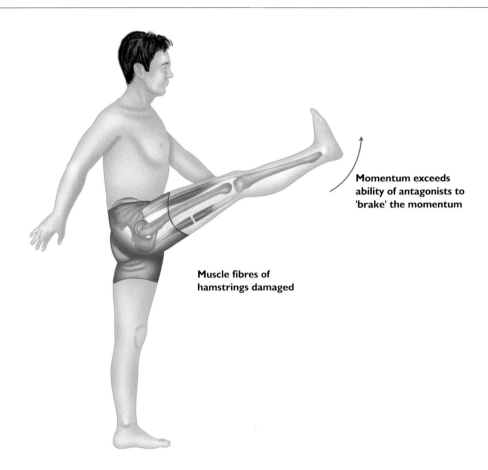

Momentum exceeds ability of antagonists to 'brake' the momentum

Muscle fibres of hamstrings damaged

Figure 156: Damage that can be caused by an over zealous ballistic stretch.

Core Stability

During day-to-day activity, skeletal muscles are acting as either stabilising muscles or muscles of movement (as outlined under Group Action of Muscles, p.136). Stabilising muscles maintain posture or hold the body in a given position as a 'platform' so that other muscles can cause the body to move in some way.

Stabilising muscles tend to be situated deep within the body. To maintain posture or a steady 'platform', their fibres perform a minimal contraction over an extended period of time. Hence they are built for endurance and therefore have many slow-twitch fibres (*see* 'red and white muscle fibres' – p.124). Persons with poor postural alignment or an inactive lifestyle tend to have insufficient tone in these muscles, which further exacerbates their poor posture and lessens their ability to stabilise functional movements.

When the stabilising muscles are under-used, the nerve impulses find it more difficult to get through to those muscles, leading to what is referred to as *poor recruitment*. That means, if we do not use a muscle for an extended period of time, we will find it more difficult to re-enervate that muscle back into use. Consequently, the majority of people in modern society would benefit from exercises that specifically address their neglected deep postural muscles.

It is particularly important to maintain your torso as a stable platform relative to the movements carried out by your limbs. As your torso or mid-section is the 'core' of your body, its success as a stable platform is referred to as *core stability*. Good core stability therefore allows you to maintain a rigid mid-section without gravity or other forces interfering with the movement you wish to perform. Core stability muscles can be retrained, especially through bracing and stabilising exercises; a fact utilised in physiotherapy treatment, Pilates, Taiji Quan, Hatha Yoga and so on. In essence, core stability can be summarised as the successful recruitment of deep muscles that maintain the natural curvatures (neutral alignment) of the spine during all other movements of the body.

Good core stability results from the deep stabilising trunk muscles co-ordinating their contraction to stabilise the spine, rather like the tightening of guide ropes around a pole or mast to give it strength and maintain its position. The deep stabilising or 'core stability' muscles collectively create what is known as an 'inner unit' of muscle. These muscles include the tranversus abdominis, multifidis, pelvic floor, diaphragm and posterior fibres of the internal oblique. The main muscles that initiate movement of the limbs whilst working in unison with the inner unit are collectively referred to as the 'outer unit' or global muscles; i.e. the spinal erectors, external and internal obliques, latissimus dorsi, gluteals, hamstrings and adductors. The following effects summarize how core stability is enhanced by some subsidiary factors of body mechanics:

Thoraco-lumbar fascia gain
As the abdominal wall is pulled in by the contraction of the transversus abdominis, the internal oblique acts synergistically to pull upon the thoraco-lumbar fascia (which wraps around the spine, connecting the deep trunk muscles to it). This in turn exerts a force on the lumbar spine that helps support and stabilise it (this force is called thoraco-lumbar gain). More specifically, the increased tension of the thoraco-lumbar fascia compresses the erector spinae and multifidis muscles, encouraging these to contract and resist the forces that are trying to flex the spine. The classic analogy is that of the guide ropes of a tent acting together to support the main structure of the tent.

Research demonstrates that in addition to the above, the paraspinal muscles – interspinalis and intertransversarii (*see* p.221–226) – assist core stability insofar as they provide an individual stabilising effect on their adjacent vertebrae, acting in a similar way to ligaments.

It is not just the recruitment of these deep-trunk muscles that is significant, but also how and when they are recruited that is important. Two key researchers in core stability theory, Hodges and Richardson, showed that co-contraction of the transversus abdominis and multifidis muscles occurs *prior* to any movement of the limbs. This suggests that these muscles anticipate dynamic forces that may act on the lumbar spine and stabilise the area before any movement takes place elsewhere.

Intra-abdominal pressure
Pressure in the abdominal cavity is increased as a result of the abdominal wall being pulled inwards by the transversus abdominis, along with a co-contraction of the pelvic floor, internal oblique and low back muscles. This in turn exerts a tensile force on the rectus sheath, which encloses the rectus abdominis muscle (*see* p.247). Because the rectus sheath attaches to the internal oblique and transversus abdominis muscles, it effectively surrounds the abdomen. The tension of the rectus sheath therefore increases the pressure within the abdomen like a pressurised balloon. This further facilitates the stability of the core. In practice we clearly experience this when we hold our breath during a significant lifting or throwing action, during which time we can feel ourselves contract the diaphragm and pelvic floor muscles.

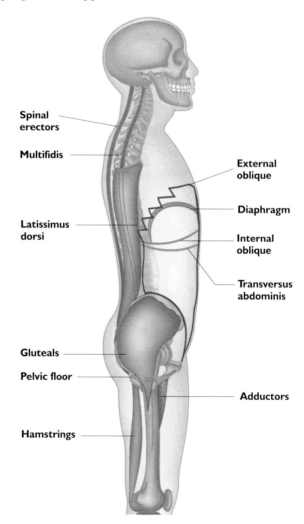

Figure 157: Schematic diagram of the core stability (inner unit) muscles and the global (outer unit) muscles. For a clear anatomical representation of each muscle (group), refer to the following pages: obliques (243, 244); diaphragm (241); transversus abdominis (246); adductors (328–330); hamstrings (325–327); gluteals (313, 315, 316); latissimus dorsi (263); mutifidis (229); spinal erectors (205–218).

The Muscular System

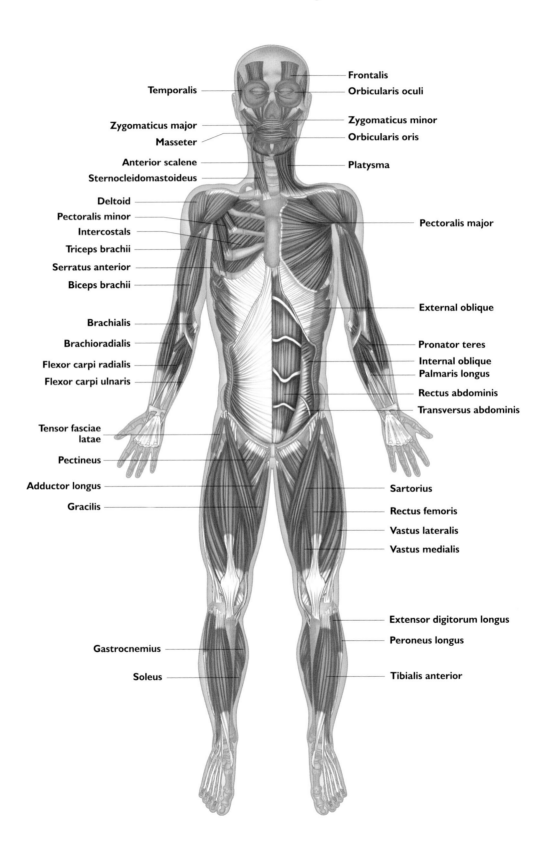

Figure 158: Muscular system (anterior view).

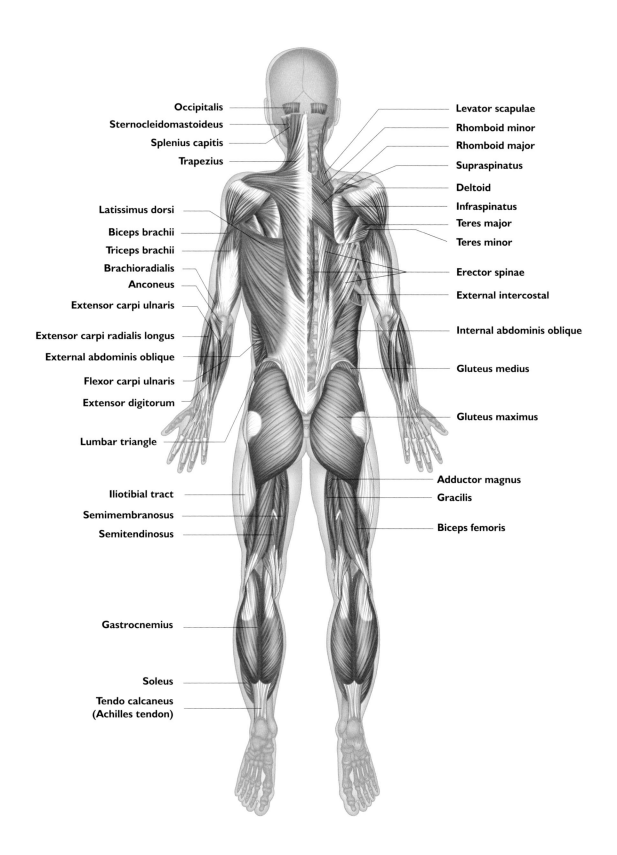

Occipitalis

Sternocleidomastoideus

Splenius capitis

Trapezius

Latissimus dorsi

Biceps brachii

Triceps brachii

Brachioradialis

Anconeus

Extensor carpi ulnaris

Extensor carpi radialis longus

External abdominis oblique

Flexor carpi ulnaris

Extensor digitorum

Lumbar triangle

Iliotibial tract

Semimembranosus

Semitendinosus

Gastrocnemius

Soleus

Tendo calcaneus
(Achilles tendon)

Levator scapulae

Rhomboid minor

Rhomboid major

Supraspinatus

Deltoid

Infraspinatus

Teres major

Teres minor

Erector spinae

External intercostal

Internal abdominis oblique

Gluteus medius

Gluteus maximus

Adductor magnus

Gracilis

Biceps femoris

Figure 159: Muscular system (posterior view).

Muscles of the Scalp and Face

8

MUSCLES OF THE SCALP

Epicranius–occipitalis

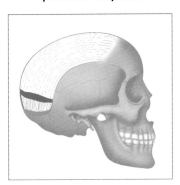

Epicranius–frontalis

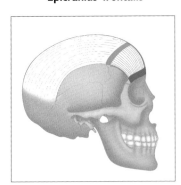

Temporoparietalis

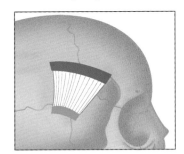

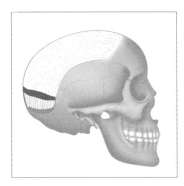

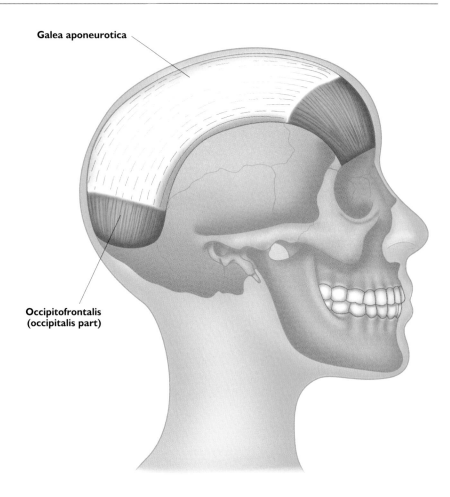

Galea aponeurotica

Occipitofrontalis
(occipitalis part)

Latin, *occiput*, back of the skull.

The epicranius (occipitofrontalis) is effectively two muscles (occipitalis and frontalis), united by an aponeurosis called the *galea aponeurotica*, so named because it forms what resembles a helmet upon the skull.

Origin
Lateral two-thirds of superior nuchal line of occipital bone. Mastoid process of temporal bone.

Insertion
Galea aponeurotica (a sheet-like tendon leading to frontal belly).

Action
Pulls scalp backward. Assists frontal belly to raise eyebrows and wrinkle forehead.

Nerve
Facial **V11** nerve (posterior auricular branch).

Artery
Occipital artery
(from external carotid artery).

Basic functional movement
Facilitates facial expressions.

EPICRANIUS–FRONTALIS

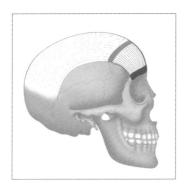

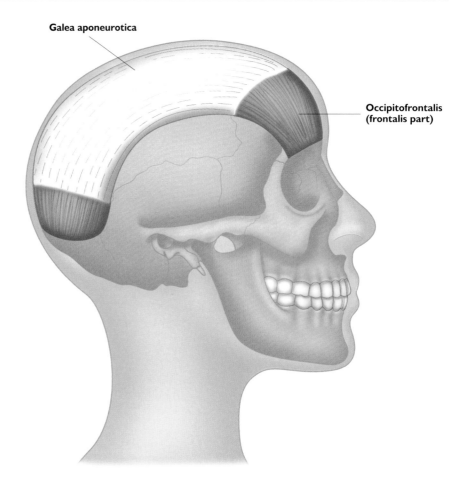

Galea aponeurotica

Occipitofrontalis
(frontalis part)

Latin, *frons*, forehead, front.

The epicranius (occipitofrontalis) is effectively two muscles (occipitalis and frontalis), united by an aponeurosis called the *galea aponeurotica*, so named because it forms what resembles a helmet upon the skull.

Origin
Galea aponeurotica.

Insertion
Fascia and skin above eyes and nose.

Action
Pulls scalp forwards. Raises eyebrows and wrinkles skin of forehead horizontally.

Nerve
Facial **V11** nerve (temporal branches).

Artery
Supraorbital and supratrochlear branches of ophthalmic artery
(from internal carotid artery).

Basic functional movement
Facilitates facial expressions.

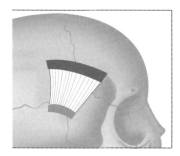

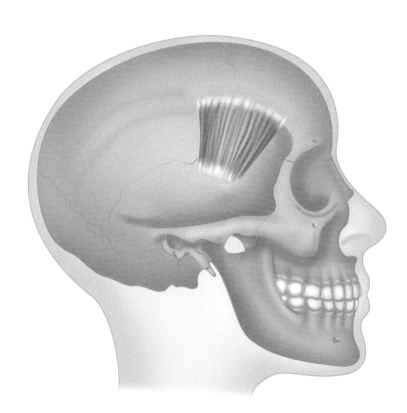

Latin, *temporis*, time, temple; *parietalis*, of the walls of a cavity.

Origin
Fascia above ear.

Insertion
Lateral border of galea aponeurotica.

Action
Tightens scalp. Raises ears.

Nerve
Facial **V11** nerve (temporal branch).

Artery
Superficial temporal and posterior auricular arteries
via external carotid artery (from common carotid artery).

MUSCLES OF THE EAR

The auricularis anterior, superior and posterior are also referred to as the extrinsic muscles of the auricle. They are generally non-functional in humans unless trained.

Auricularis anterior

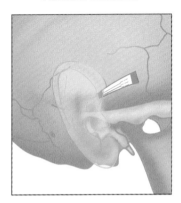

Auricularis superior

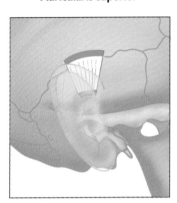

Auricularis posterior

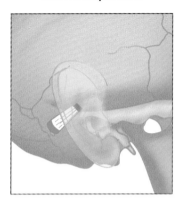

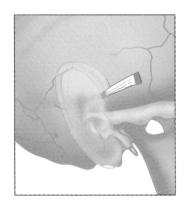

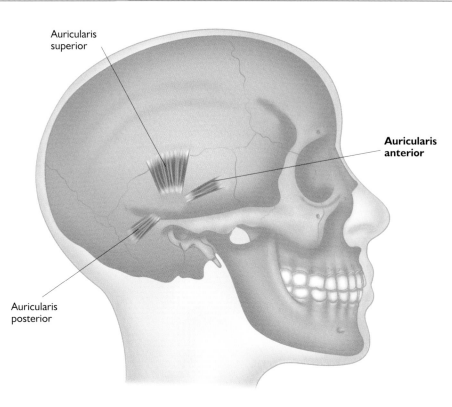

Auricularis
superior

Auricularis
anterior

Auricularis
posterior

Latin, *auricularis*, pertaining to the ear; *anterior*, before.

Origin
Fascia in temporal region anterior to ear.

Insertion
Anterior to helix of ear.

Action
Draws ear forward. Moves scalp.

Nerve
Facial **V11** nerve (temporal branch).

Artery
Superficial temporal and posterior auricular arteries
via external carotid artery (from common carotid artery).

AURICULARIS SUPERIOR / POSTERIOR

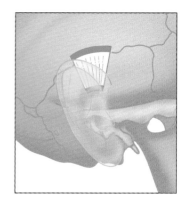

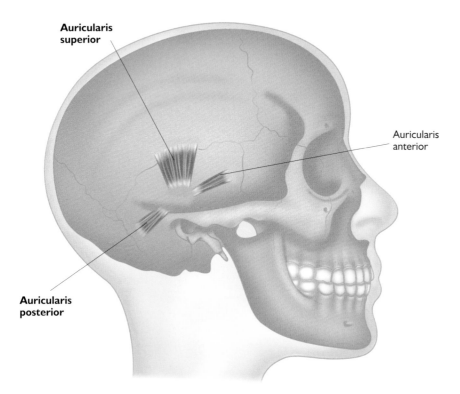

Auricularis superior

Auricularis anterior

Auricularis posterior

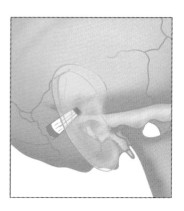

Latin, *auricularis*, pertaining to the ear; *superior*, above.

Origin
Fascia in temporal region above ear.

Insertion
Superior part of ear.

Action
Draws ear forward. Moves scalp.

Nerve
Facial **V11** nerve (temporal branch).

Artery
Superficial temporal and posterior auricular arteries
via external carotid artery (from common carotid artery).

Latin, *auricularis*, pertaining to the ear; *posterior*, behind.

Origin
Temporal bone, near mastoid process.

Insertion
Posterior part of ear.

Action
Pulls ear upward.

Nerve
Facial **V11** nerve (posterior auricular branch).

Artery
Superficial temporal and posterior auricular arteries
via external carotid artery (from common carotid artery).

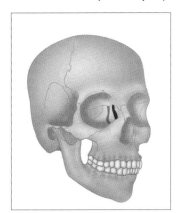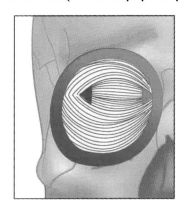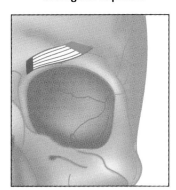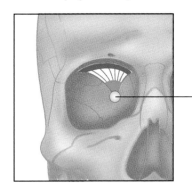

Muscles of the Scalp and Face

MUSCLES OF THE EYELIDS

Orbicularis oculi (lacrimal part)

Orbicularis oculi (orbital and palpebral part)

Corrugator supercilii

Levator palpebrae superioris

Optic nerve

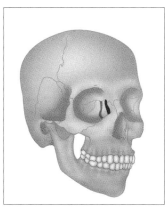

Lacrimal part

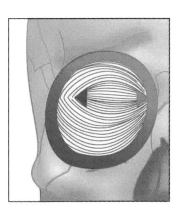

Orbital and palpebral part

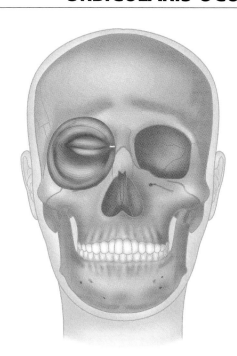

Latin, *orbis*, orb, circle; *oculi*, of the eye.

This complex and extremely important muscle consists of three parts, which together form an important protective mechanism surrounding the eye.

ORBITAL PART

Origin
Frontal bone. Medial wall of orbit (on maxilla).

Insertion
Circular path around orbit, returning to origin.

Action
Strongly closes eyelids (firmly 'screws up' the eye).

Nerve
Facial **V11** nerve (temporal and zygomatic branches).

Artery
Upper fibres: Supraorbital and supratrochlear branches of ophthalmic artery (from internal carotid artery).
Lower fibres: Infraorbital branch of maxillary artery and angular branch of facial artery (from external carotid artery).

PALPEBRAL PART (in eyelids)

Latin, pertaining to an eyelid.

Origin
Medial palpebral ligament.

Insertion
Lateral palpebral ligament into zygomatic bone.

Action
Gently closes eyelids (and comes into action involuntarily, as in blinking).

Nerve
Facial **V11** nerve (temporal and zygomatic branches).

Artery
Upper fibres: Supraorbital and supratrochlear branches of ophthalmic artery (from internal carotid artery).
Lower fibres: Infraorbital branch of maxillary artery and angular branch of facial artery (from external carotid artery).

LACRIMAL PART (behind medial palpebral ligament and lacrimal sac)

Latin, pertaining to the tears.

Origin
Lacrimal bone.

Insertion
Lateral palpebral raphe.

Action
Dilates lacrimal sac and brings lacrimal canals onto surface of eye.

Nerve
Facial **V11** nerve (temporal and zygomatic branches).

Artery
Infraorbital branch of maxillary artery and angular branch of facial artery (from external carotid artery).

LEVATOR PALPEBRAE SUPERIORIS

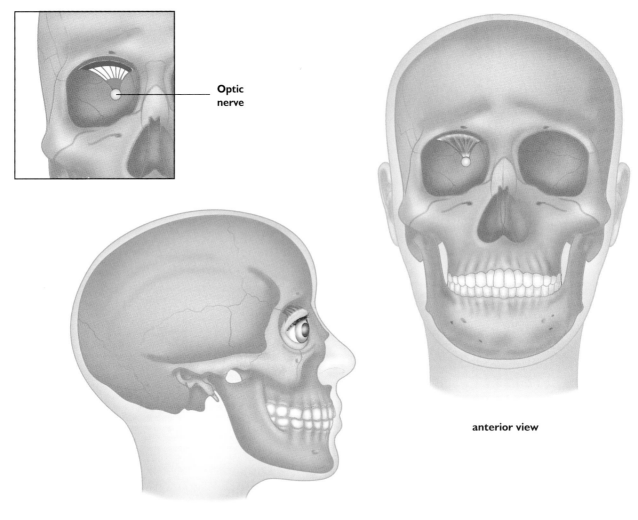

Optic
nerve

anterior view

lateral view

Latin, *levare*, to raise; *palpebral*, pertaining to an eyelid; *superior*, above.

This muscle is unusual in that it contains both somatic and visceral muscle fibres. It is the antagonist of the palpebral part of the orbicularis oculi. Therefore, paralysis of the levator results in the upper eyelid drooping down over the eyeball.

Origin
Root of orbit (lesser wing of sphenoid bone).

Insertion
Skin of upper eyelid.

Action
Raises upper eyelid.

Nerve
Oculomotor 111 nerve.

Artery
Ophthalmic artery
(from internal carotid artery).

Basic functional movement
Waking up.

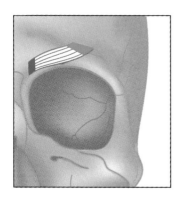

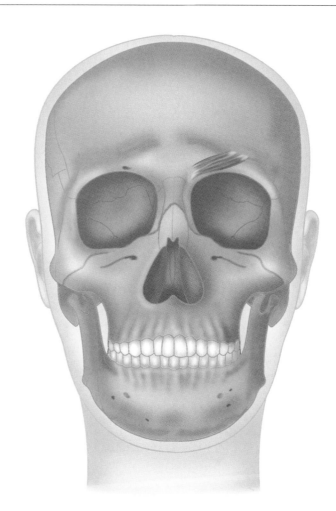

Latin, *corrugator*, muscle which wrinkles; *supercilii*, of the eyebrow.

Origin
Medial end of superciliary arch of frontal bone.

Insertion
Deep surface of skin under medial half of the eyebrows.

Action
Draws eyebrows medially and downward, so producing vertical wrinkles, as in frowning.

Nerve
Facial **V11** nerve (temporal branch).

Artery
Supratrochlear branch of ophthalmic artery
(from internal carotid artery).

Basic functional movement
Facilitates facial expression.

MUSCLES OF THE NOSE

Procerus

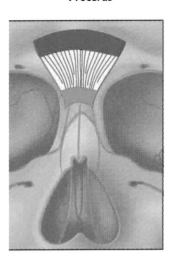

Nasalis

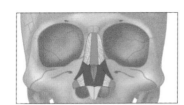

Depressor septi nasi

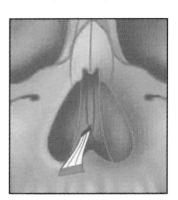

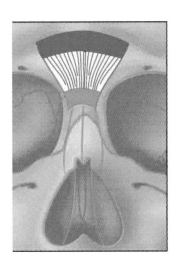

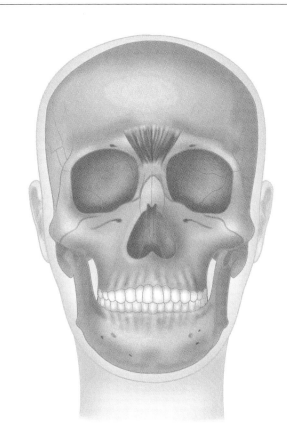

Latin, long, slender.

Origin
Fascia over nasal bone. Lateral nasal cartilage.

Insertion
Skin between eyebrows.

Action
Wrinkles nose. Pulls medial portion of eyebrows downwards.

Nerve
Facial **V11** nerve.

Artery
Supratrochlear branch of ophthalmic artery
(from internal carotid artery).

Basic functional movement
Example: Enables strong 'sniffing' and sneezing.

NASALIS

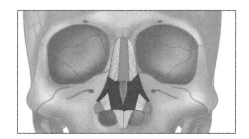

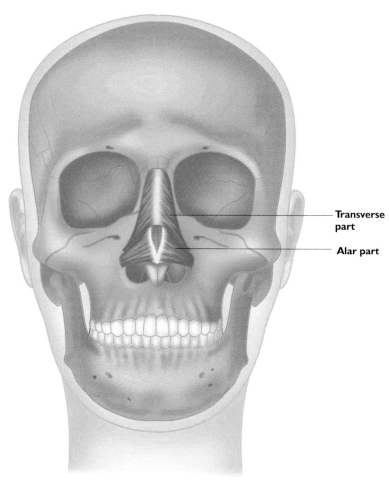

Transverse part

Alar part

Latin, *nasus*, nose.

Origin
Middle of maxilla (above incisor and canine teeth). Greater alar cartilage. Skin on nose.

Insertion
Joins muscle of opposite side across bridge of nose. Skin at tip of nose.

Action
Maintains opening of external nares during forceful inhalation (i.e. flares the nostrils).

Nerve
Facial **V11** nerve (buccal branches).

Artery
Superior labial branch of the facial artery
(from external carotid artery).

Basic functional movement
Example: Strongly breathing in through the nose.

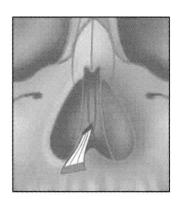

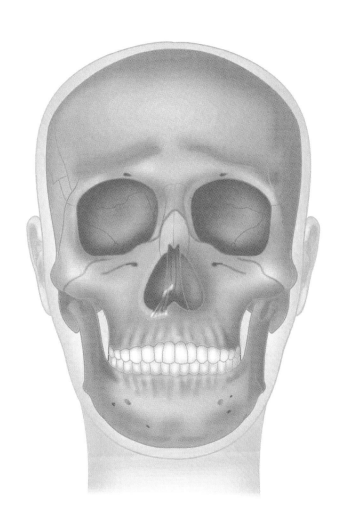

Latin, *deprimere*, to press down; *septum*, dividing wall; *nasi*, of the nose.

Origin
Incisive fossa of maxilla (above incisor teeth).

Insertion
Nasal septum and ala.

Action
Constricts nares.

Nerve
Facial **V11** nerve (buccal branches).

Artery
Superior labial branch of the facial artery
(from external carotid artery).

Basic functional movement
Example: Twitching the nose.

MUSCLES OF THE MOUTH

Orbicularis oris

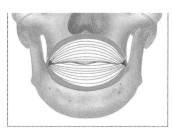

Zygomaticus minor

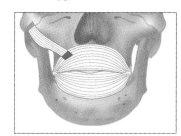

Risorius

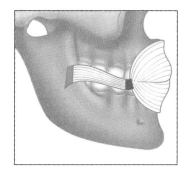

Levator labii superioris

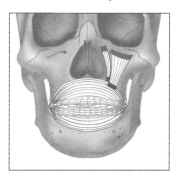

Depressor labii inferioris

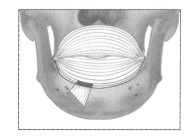

Platysma

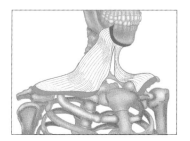

Levator anguli oris

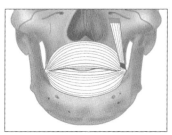

Depressor anguli oris

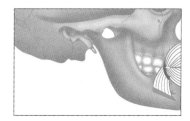

Buccinator

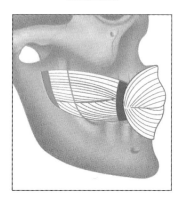

Zygomaticus major

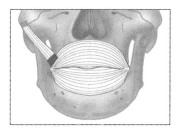

Mentalis

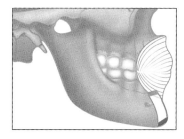

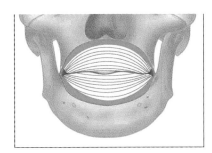

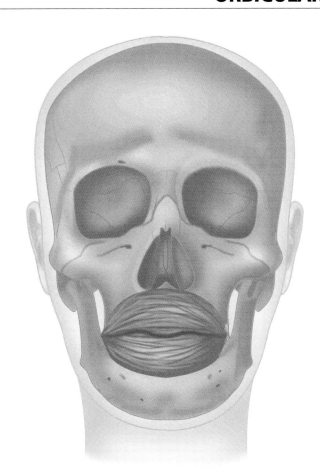

Latin, *orbis*, orb, circle; *oris*, pertaining to the mouth.

This is a composite sphincter muscle that encircles the mouth. It receives fasciculi from many other muscles.

Origin
Muscle fibres surrounding the opening of mouth, attached to the skin, muscle and fascia of the lips and surrounding area.

Insertion
Skin and fascia at corner of mouth.

Action
Closes lips, compresses lips against teeth, protrudes (purses) lips, and shapes lips during speech.

Nerve
Facial **V11** nerve (buccal and mandibular branches).

Artery
Superior and inferior labial branches of the facial artery
(from external carotid artery).

Basic functional movement
Facial expressions involving the lips.

LEVATOR LABII SUPERIORIS

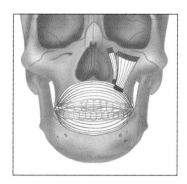

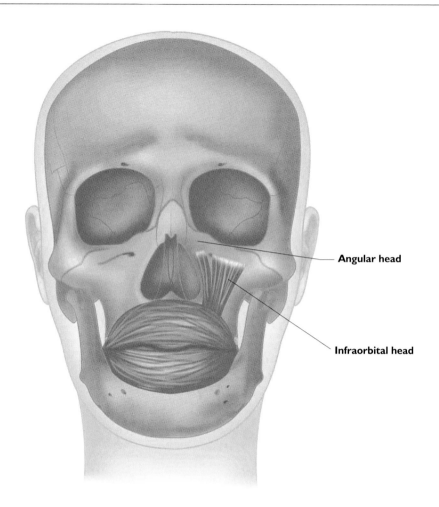

Angular head

Infraorbital head

Latin, *levare*, to raise; *labium*, lip; *superioris*, above.

ANGULAR HEAD

Origin
Zygomatic bone and frontal process of maxilla.

Insertion
Greater alar cartilage, upper lip and skin of nose.

Action
Raises upper lip. Dilates nares. Forms nasolabial furrow.

Nerve
Facial **V11** nerve (buccal branches).

Artery
Infraorbital artery
(via maxillary artery, from external carotid artery).

Basic functional movement
Facilitates facial expression and kissing.

INFRAORBITAL HEAD

Origin
Lower border of orbit.

Insertion
Muscles of upper lip.

Action
Raises upper lip.

Nerve
Facial **V11** nerve (buccal branches).

Artery
Superior labial branch of the facial artery
(from external carotid artery).

Basic functional movement
Facilitates facial expression and kissing.

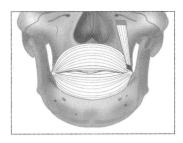

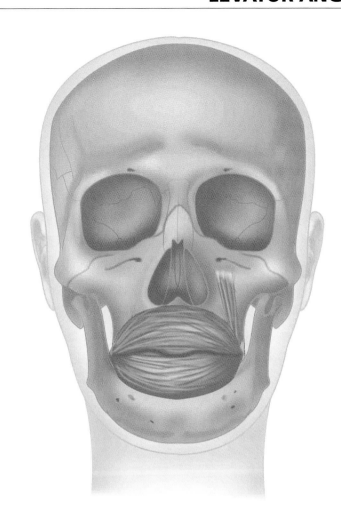

Latin, *levare*, to raise; *anguli*, triangular area; *oris*, pertaining to the mouth.

Origin
Canine fossa of maxilla.

Insertion
Corner of mouth.

Action
Elevates angle (corner) of mouth.

Nerve
Facial **V11** nerve (buccal branches).

Artery
Superior labial branch of the facial artery
(from external carotid artery).

Basic functional movement
Helps produce a smiling expression.

ZYGOMATICUS MAJOR

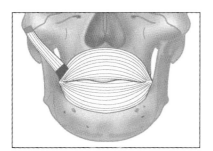

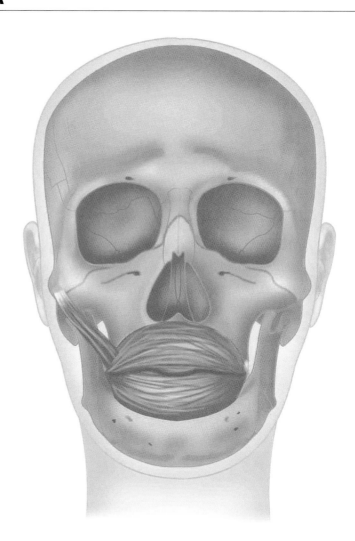

Greek, *zygon*, yoke, union; **Latin**, *major*, large.

Origin
Upper lateral surface of zygomatic bone.

Insertion
Skin at corner of mouth. Orbicularis oris.

Action
Pulls corner of mouth up and back, as in smiling.

Nerve
Facial **V11** nerve (zygomatic and buccal branches).

Artery
Transverse facial artery and facial artery
(from external carotid artery).

Basic functional movement
Smiling.

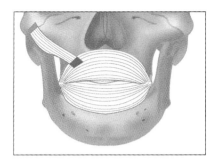

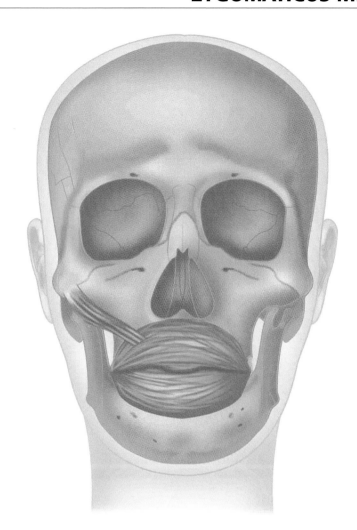

Greek, *zygon*, yoke, union; **Latin**, *minor*, small.

Origin
Lower surface of zygomatic bone.

Insertion
Lateral part of upper lip lateral to levator labii superioris.

Action
Elevates the upper lip. Forms nasolabial furrow.

Nerve
Facial **V11** nerve (buccal branches).

Artery
Transverse facial artery and facial artery
(from external carotid artery).

Basic functional movement
Facilitates facial expression.

DEPRESSOR LABII INFERIORIS

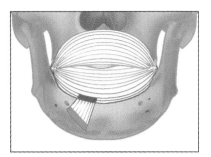

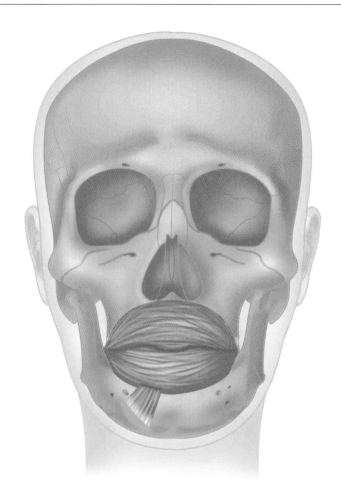

Latin, *deprimere,* to press down; *labii,* of the lip; *inferior,* below.

Origin
Anterior surface of mandible, between mental foramen and symphysis.

Insertion
Skin of lower lip.

Action
Pulls lower lip downward and slightly laterally.

Nerve
Facial **V11** nerve (marginal mandibular branch).

Artery
Inferior labial and submental branches of the facial artery
(from external carotid artery).

Basic functional movement
Facilitates facial expression.

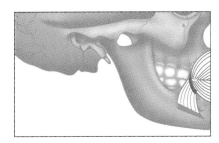

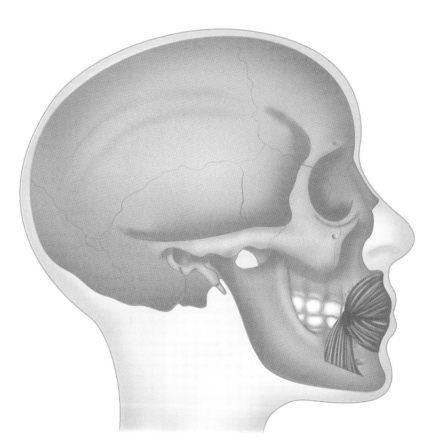

Latin, *deprimere*, to press down; *angulus*, angle; *oris*, pertaining to the mouth.

Muscle fibres are continuous with the platysma.

Origin
Oblique line of the mandible.

Insertion
Corner of mouth.

Action
Pulls corner of mouth downwards, as in sadness or frowning.

Nerve
Facial **V11** nerve (marginal mandibular and buccal branches).

Artery
Inferior labial and submental branches of the facial artery
(from external carotid artery).

MENTALIS

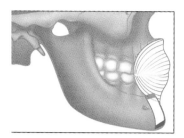

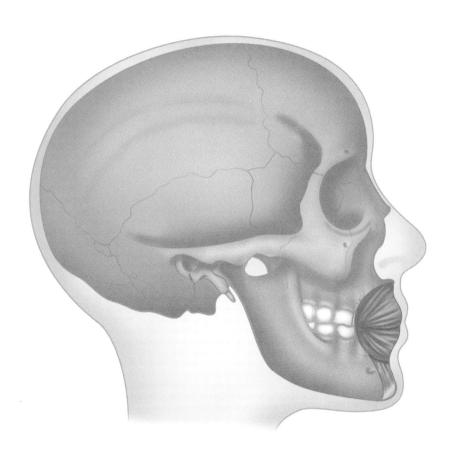

Latin, relating to the chin.

This is the only muscle of the lips that normally has no connection with the orbicularis oris.

Origin
Incisive fossa of anterior surface of mandible.

Insertion
Skin of chin.

Action
Protrudes lower lip and pulls up (wrinkles) skin of chin, as in pouting.

Nerve
Facial **V11** nerve (marginal mandibular branch).

Artery
Inferior labial and submental branches of the facial artery
(from external carotid artery).

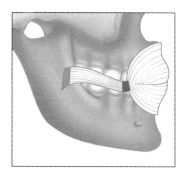

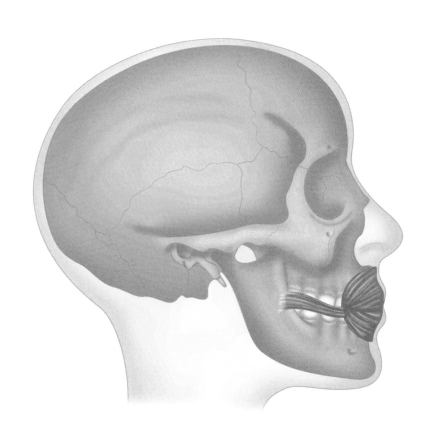

Latin, *risus*, laughter.

This thin muscle is often completely fused with the platysma.

Origin
Fascia over masseter and parotid (salivary) gland (i.e. fascia of the lateral cheek).

Insertion
Skin at angle of mouth.

Action
Draws angle of mouth laterally, as in tenseness or grinning.

Nerve
Facial **V11** nerve (buccal branches).

Artery
Transverse facial artery and facial artery
(from external carotid artery).

PLATYSMA

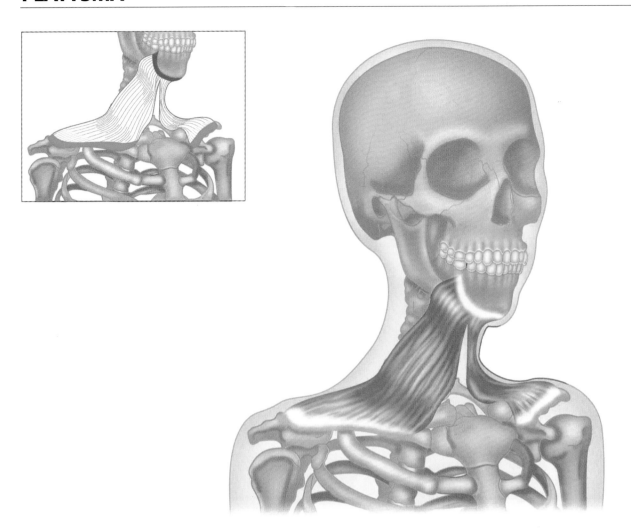

Greek, *platy,* broad, flat.

This muscle may be seen to stand out in a runner finishing a hard race.

Origin
Subcutaneous fascia of upper quarter of chest (i.e. fascia overlying the pectoralis major and deltoid muscles).

Insertion
Subcutaneous fascia and muscles of chin and jaw. Inferior border of mandible.

Action
Pulls lower lip from corner of mouth downwards and laterally. Draws skin of chest upwards.

Nerve
Facial **V11** nerve (cervical branch).

Artery
Facial artery
(from external carotid artery).

Basic functional movement
Example: Gives expression of being startled or of sudden fright.

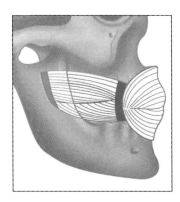

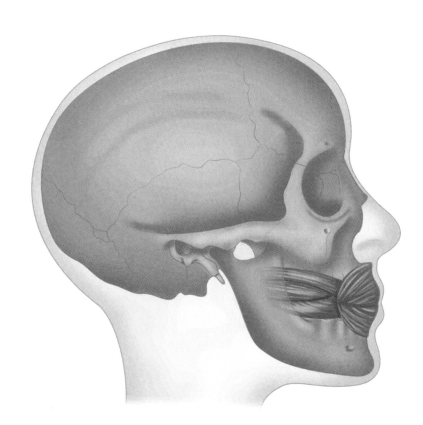

Latin, *buccina*, trumpet; *bucca*, cheek.

This muscle forms the substance of the cheek.

Origin
Alveolar processes of maxilla and mandible over molars and along pterygomandibular raphe (fibrous band extending from the pterygoid hamulus to the mandible).

Insertion
Orbicularis oris (muscles of lips).

Action
Compresses cheek as in blowing air out of mouth, and caves cheeks in, producing the action of sucking.

Nerve
Facial **V11** nerve (buccal branches).

Artery
Facial artery
(from external carotid artery).

MUSCLES OF MASTICATION

Masseter

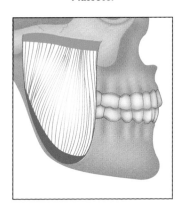

Pterygoideus lateralis

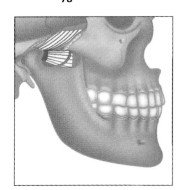

Pterygoideus medialis

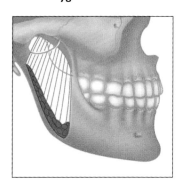

Temporalis

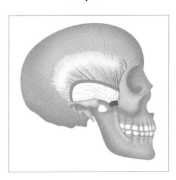

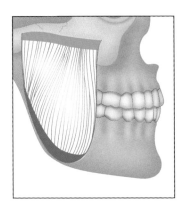

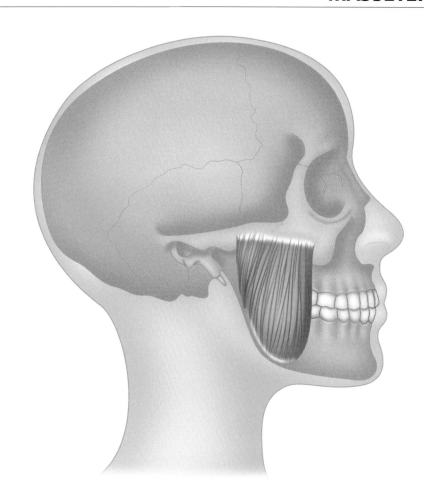

Greek, *maseter*, chewer.

The masseter is the most superficial muscle of mastication, easily felt when the jaw is clenched.

Origin
Zygomatic process of maxilla. Medial and inferior surfaces of zygomatic arch.

Insertion
Angle of ramus of mandible. Coronoid process of mandible.

Action
Closes jaw. Clenches teeth. Assists in side to side movement of mandible.

Nerve
Trigeminal **V** nerve (mandibular division).

Artery
Masseteric branch of the maxillary artery
(from external carotid artery).

Basic functional movement
Chewing food.

TEMPORALIS

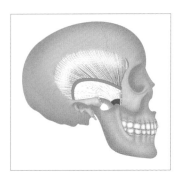

Zygomatic arch
has been removed.

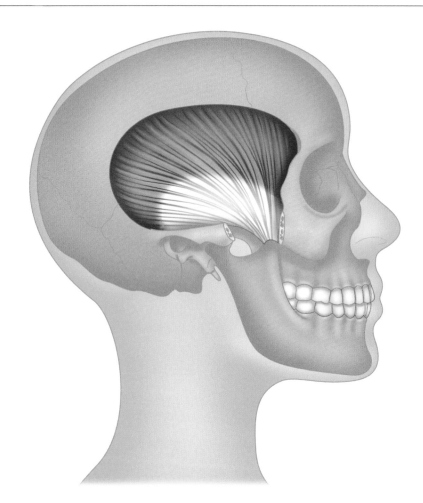

Latin, pertaining to the lateral side of the head, time.

Origin
Temporal fossa, including parietal, temporal and frontal bones. Temporal fascia.

Insertion
Coronoid process of mandible. Anterior border of ramus of mandible.

Action
Closes jaw. Clenches teeth. Assists in side to side movement of mandible.

Nerve
Anterior and posterior deep temporal nerves from the trigeminal **V** nerve (mandibular division).

Artery
Anterior and posterior deep temporal branches of maxillary artery
(from external carotid artery).

Basic functional movement
Chewing food.

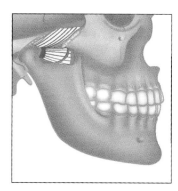

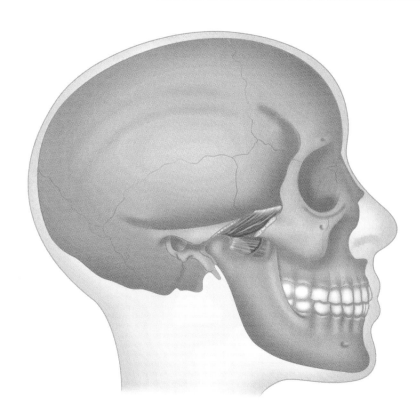

Greek, *pterygodes*, like a wing; **Latin**, *lateral*, to the side.

The superior head of this muscle is sometimes called sphenomeniscus, because it inserts into the disc of the temporomandibular joint.

Origin
Superior head: lateral surface of greater wing of sphenoid.
Inferior head: lateral surface of lateral pterygoid plate of sphenoid.

Insertion
Superior head: capsule and articular disc of the temporomandibular joint.
Inferior head: neck of mandible.

Action
Protrudes mandible. Opens mouth. Moves mandible from side to side (as in chewing).

Nerve
Trigeminal **V** nerve (mandibular division).

Artery
Lateral pterygoid branch of the maxillary artery
(from external carotid artery).

Basic functional movement
Chewing food.

PTERYGOIDEUS MEDIALIS (Medial Pterygoid)

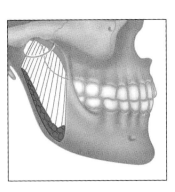

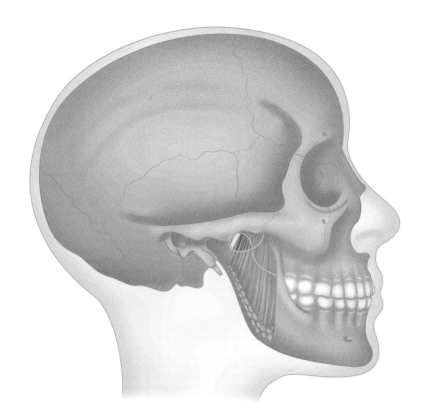

Greek, *pterygodes*, like a wing; **Latin**, *medius*, middle.

This muscle mirrors the masseter muscle in both its position and action, with the ramus of the mandible positioned between the two muscles.

Origin
Medial surface of lateral pterygoid plate of the sphenoid bone. Pyramidal process of the palatine bone. Tuberosity of maxilla.

Insertion
Medial surface of the ramus and the angle of the mandible.

Action
Elevates and protrudes the mandible. Therefore it closes the jaw and assists in side to side movement of the mandible, as in chewing.

Nerve
Trigeminal **V** nerve (mandibular division).

Artery
Medial pterygoid branch of the maxillary artery
(from external carotid artery).

Basic functional movement
Chewing food.

Muscles of the Neck

9

HYOID MUSCLES

The hyoid muscles are mostly concerned with steadying or moving the hyoid bone, and therefore the tongue and larynx which are attached to it.

Mylohyoideus

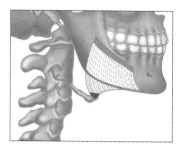

Sternohyoideus

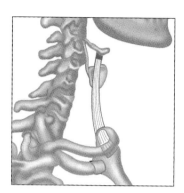

Omohyoideus

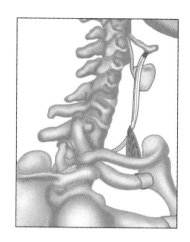

Geniohyoideus

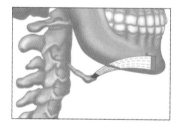

Sternothyroideus

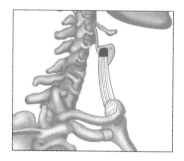

Thyrohyoideus

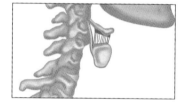

Stylohyoideus

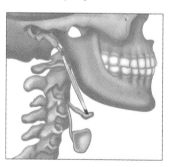

Digastricus

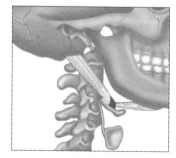

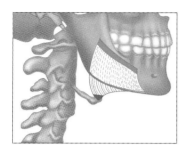

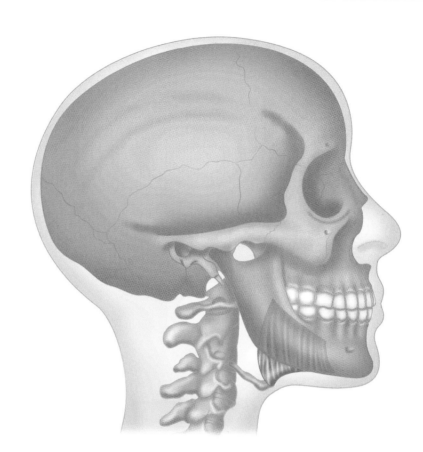

Greek, *myloi*, pertaining to the molar teeth; *hyoid*, hyoeides, shaped like the Greek letter upsilon (ν).

The mylohyoid fibres form a sling or diaphragm that supports the floor of the mouth.

Origin
Mylohyoid line on the inner surface of mandible.

Insertion
Hyoid bone.

Action
Raises floor of mouth in swallowing. Elevates hyoid bone. Helps press tongue upwards and backwards against roof of mouth.

Nerve
Mylohyoid nerve from the inferior alveolar nerve, which is a branch of the trigeminal **V** nerve (mandibular division).

Artery
Mylohyoid branch of the inferior alveolar branch of the maxillary artery
(from external carotid artery).

Basic functional movement
Swallowing.

GENIOHYOIDEUS

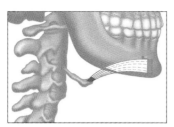

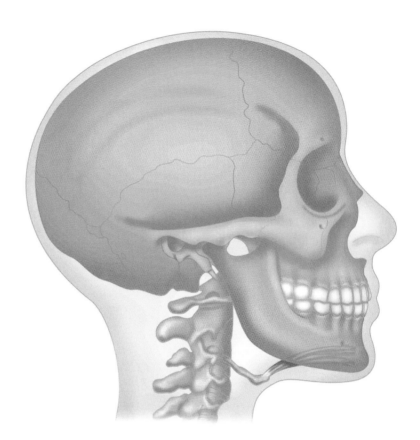

Greek, *geneion*, chin; *hyoid*, hyoeides, shaped like the Greek letter upsilon (ν).

Origin
Lower part of mental spine of interior medial surface of mandible.

Insertion
Hyoid bone.

Action
Protrudes and elevates hyoid bone, widening the pharynx for the reception of food. Can help retract and depress the mandible if the hyoid bone is fixed.

Nerve
Fibres of cervical nerve C1, conveyed by the hypoglossal nerve **X11**.

Artery
Lingual artery, and submental branch of the facial artery
(from external carotid artery).

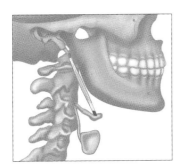

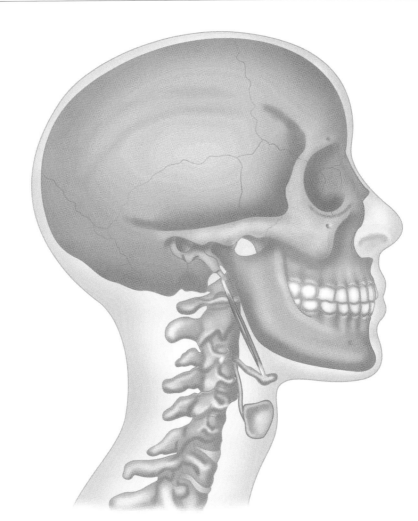

Latin, *stylus*, stake, pole; *hyoid*, hyoeides, shaped like the Greek letter upsilon (ν).

Origin
Posterior border of styloid process of temporal bone.

Insertion
Hyoid bone (after splitting to enclose the intermediate tendon of the digastric muscle).

Action
Pulls hyoid bone upward and backward, thereby elevating the tongue.

Nerve
Facial **V11** nerve.

Artery
Ascending pharyngeal artery, and may receive supply from branches of facial artery (from external carotid artery).
Can also receive supply from branches of facial artery.

DIGASTRICUS

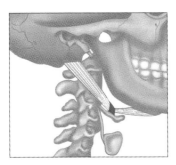

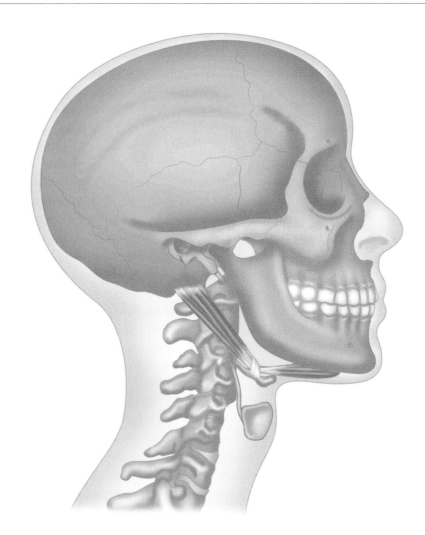

Latin, having two bellies.

Origin
Anterior belly: digastric fossa on inner side of lower border of mandible, near symphysis.
Posterior belly: mastoid notch of temporal bone.

Insertion
Body of hyoid bone via a fascial sling over an intermediate tendon.

Action
Raises hyoid bone. Depresses and retracts mandible as in opening the mouth.

Nerve
Anterior belly: mylohyoid nerve, from trigeminal **V** nerve (mandibular division).
Posterior belly: facial (**V11**) nerve.

Artery
Auricular, occipital and stylomastoid branches of the posterior auricular artery
(from external carotid artery).

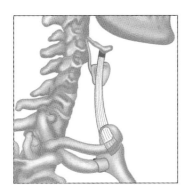

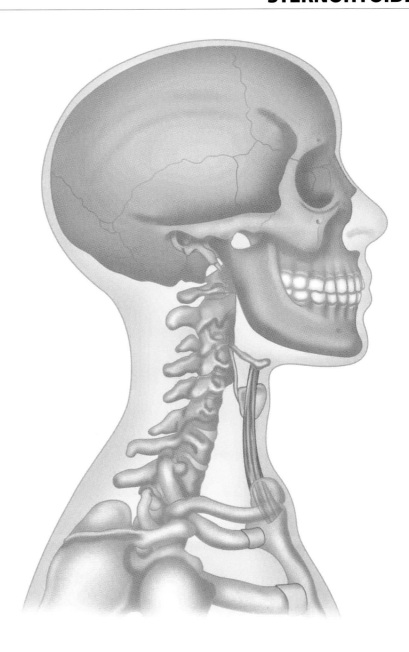

Greek, *sternon*, relating to the sternum; *hyoid*, hyoeides, shaped like the Greek letter upsilon (ν).

Origin
Posterior surface of manubrium of sternum. Medial end of clavicle.

Insertion
Lower border of hyoid bone (medial to insertion of omohyoid).

Action
Depresses hyoid bone. Stabilizes hyoid bone when other muscles are acting from it.

Nerve
Ansa cervicalis nerve, C1, **2**, **3**.

Artery
Superior thyroid artery
(from external carotid artery).
Can also receive supply from inferior thyroid artery (from thyrocervical trunk of the subclavian artery).

STERNOTHYROIDEUS

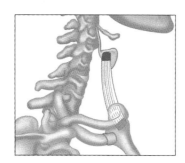

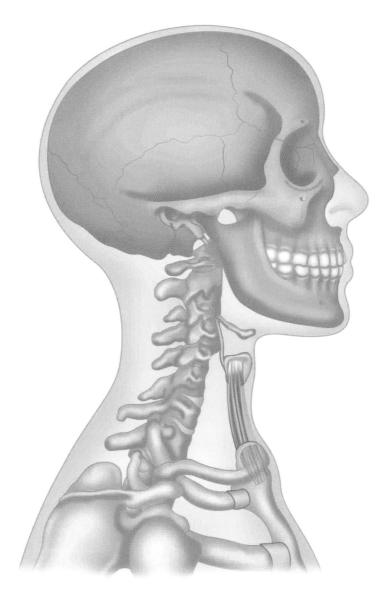

Greek, *sternon*, relating to the sternum; *thyreos*, oblong shield.

Lies deep to sternohyoideus.

Origin
Posterior surface of manubrium of sternum, below origin of sternohyoid. First costal cartilage.

Insertion
Oblique line on outer surface of thyroid cartilage.

Action
Pulls thyroid cartilage away from hyoid bone, so opening the laryngeal orifice.

Nerve
Ansa cervicalis nerve, C**1**, **2**, **3**.

Artery
Superior thyroid artery
(from external carotid artery).
Can also receive supply from inferior thyroid artery (from thyrocervical trunk of the subclavian artery).

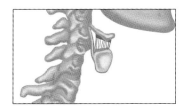

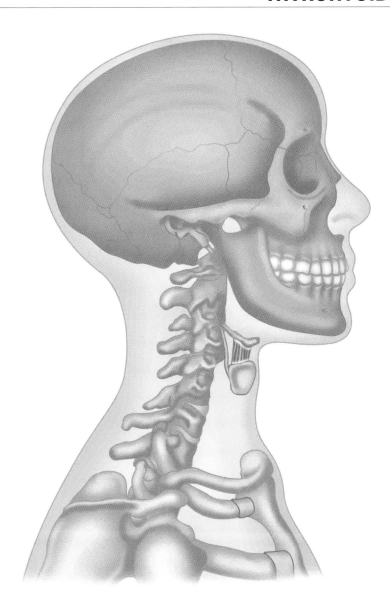

Greek, *thyreos*, oblong shield; *hyoid*, hyoeides, shaped like the Greek letter upsilon (ν).

This is a short strap muscle.

Origin
Oblique line of outer surface of thyroid cartilage.

Insertion
Lower border of body and greater horn of hyoid bone.

Action
Raises thyroid and depresses hyoid bone, thus closing laryngeal orifice, preventing food from entering the larynx during swallowing.

Nerve
Ansa cervicalis nerve, C1, 2, via fibres from the descending hypoglossal X11 nerve.

Artery
Superior thyroid artery
(from external carotid artery).
Can also receive supply from inferior thyroid artery (from thyrocervical trunk of the subclavian artery).

OMOHYOIDEUS

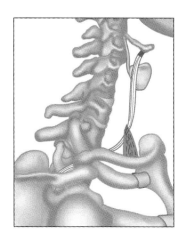

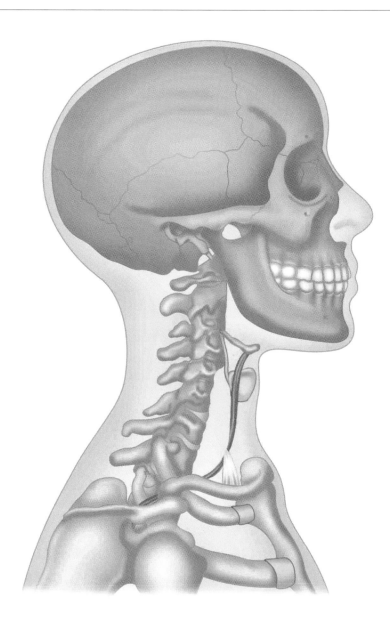

Greek, *omos*, shoulder; *hyoid*, hyoeides, shaped like the Greek letter upsilon (ν).

Origin
Inferior belly: upper border of scapula medial to the scapular notch. Superior transverse ligament.
Superior belly: intermediate tendon.

Insertion
Inferior belly: intermediate tendon.
Superior belly: lower border of hyoid bone, lateral to the insertion of sternohyoid.

NOTE: The intermediate tendon is tied down to the clavicle and first rib by a sling of the cervical fascia.

Action
Depresses hyoid bone.

Nerve
Ansa cervicalis nerve, C2, 3.

Artery
Transverse cervical artery
(from subclavian artery).
Can also receive supply from inferior thyroid artery (from thyrocervical trunk of the subclavian artery).

The anterior vertebral muscles are a small group of muscles attached to the bodies and transverse processes of the cervical and upper thoracic regions of the vertebral column.

Longus colli

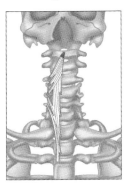

Superior oblique

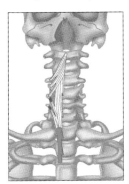

Inferior oblique

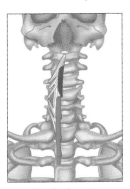

Vertical part

Longus capitis

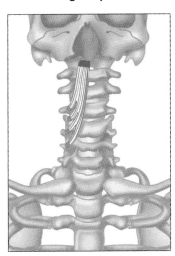

Rectus capitis anterior

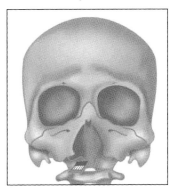

Rectus capitis lateralis

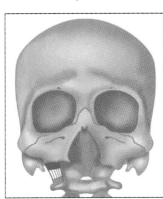

LONGUS COLLI

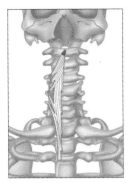

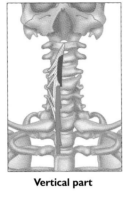

Superior oblique

Vertical part

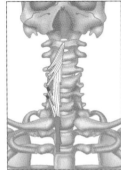

Inferior oblique

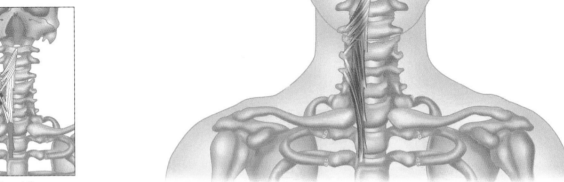

Latin, *longus*, long; *colli*, of the neck.

Longus colli, which can be divided into three parts, is the largest member of the anterior vertebral muscle group.

SUPERIOR OBLIQUE PART

Origin
Transverse processes of third, fourth and fifth cervical vertebrae.

Insertion
Anterior arch of atlas.

Action
Flexes cervical vertebrae.

Nerve
Ventral rami of cervical nerves C2–C7.

Artery
Deep cervical artery of costocervical trunk (from subclavian artery).

Basic functional movement
Gives smoothness and stability to flexion at the neck.

INFERIOR OBLIQUE PART

Origin
Anterior surface of first two or three cervical vertebral bodies.

Insertion
Transverse processes of fifth and sixth cervical vertebrae.

Action
Flexes cervical vertebrae.

Nerve
Ventral rami of cervical nerves C2–C7.

Artery
Deep cervical artery of costocervical trunk (from subclavian artery).

Basic functional movement
Gives smoothness and stability to flexion at the neck.

VERTICAL PART

Origin
Anterior surface of upper three thoracic and lower three cervical vertebral bodies.

Insertion
Transverse processes of fifth and sixth cervical vertebrae.

Action
Flexes cervical vertebrae.

Nerve
Ventral rami of cervical nerves C2–C7.

Artery
Deep cervical artery of costocervical trunk (from subclavian artery).

Basic functional movement
Gives smoothness and stability to flexion at the neck.

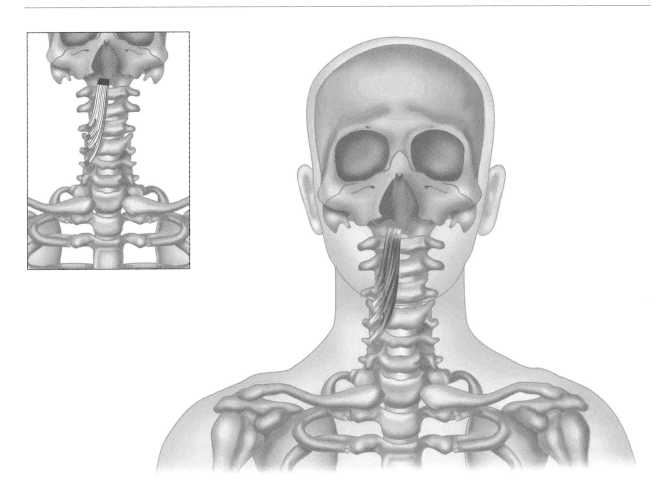

Latin, *longus*, long; *capitis*, of the head.

Longus capitis lies anterior to the superior oblique fibres of longus colli.

Origin
Transverse processes of third through to sixth cervical vertebrae.

Insertion
Occipital bone anterior to foramen magnum.

Action
Flexes head and upper part of cervical spine.

Nerve
Ventral rami of cervical nerves C1–C3 (C4).

Artery
Deep cervical artery of costocervical trunk
(from subclavian artery).

Basic functional movement
Gives smoothness and stability to flexion at head (nodding).

RECTUS CAPITIS ANTERIOR

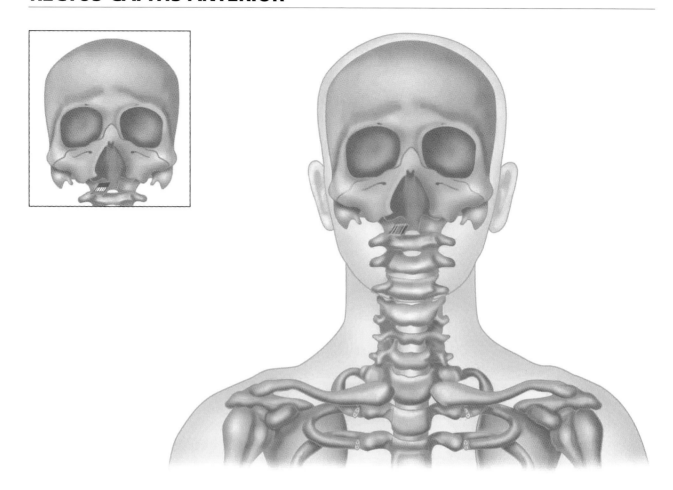

Latin, *rectum*, straight; *capitis*, of the head; *anterior*, before.

Origin
Anterior surface of the lateral mass of atlas.

Insertion
Basilar part of occipital bone anterior to the occipital condyle (i.e. between occipital condyle and longus capitis).

Action
Flexes head upon neck.
Holds articular surfaces of atlanto-occipital joint in close apposition during movements.

Nerve
Loop between ventral rami of cervical nerves C1, 2.

Artery
Deep cervical artery of costocervical trunk
(from subclavian artery).

Basic functional movement
Gives smoothness and stability to flexion at head (nodding).

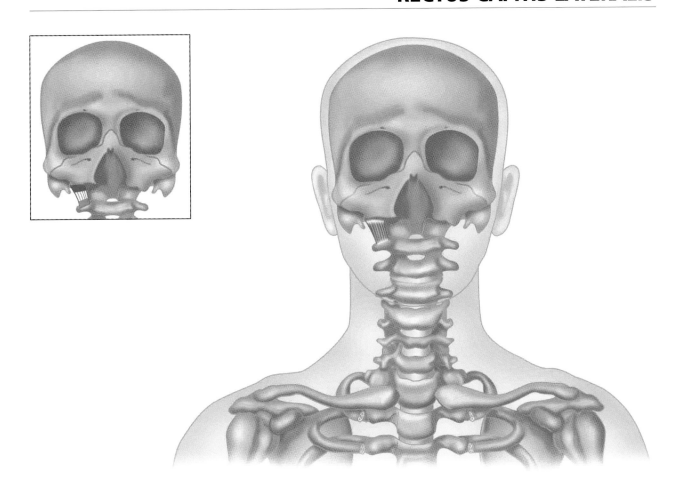

Latin, *rectum*, straight; *capitis*, of the head; *lateral*, to the side.

Origin
Transverse process of atlas.

Insertion
Jugular process of occipital bone.

Action
Tilts head laterally to same side.
Stabilizes atlanto-occipital joint.

Nerve
Loop between ventral rami of cervical nerves C1, 2.

Artery
Deep cervical artery of costocervical trunk
(from subclavian artery).

LATERAL VERTEBRAL MUSCLES

The lateral vertebral muscles of the neck comprise the scalene group, which run from the transverse processes of the cervical vertebrae downwards to the ribs, plus the sternocleidomastoideus.

Scalenus anterior

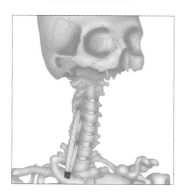

Scalenus posterior

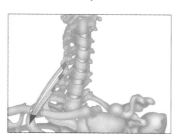

Sternocleidomastoideus

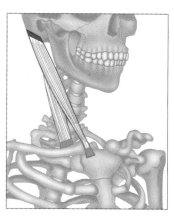

Scalenus medius

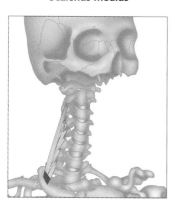

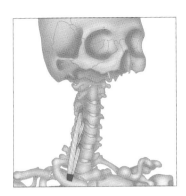

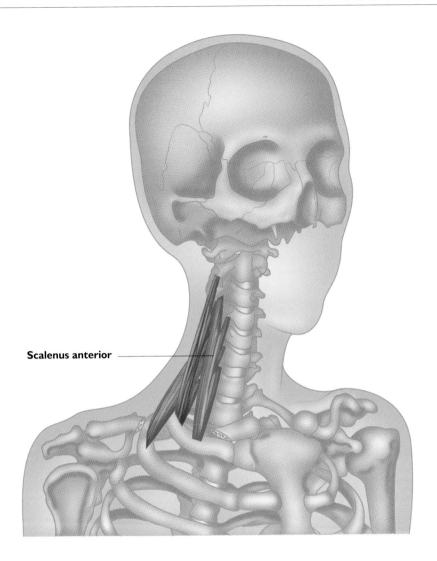

Scalenus anterior

Greek, *skalenos*, uneven; **Latin**, *anterior*, before.

Origin
Transverse processes of third through to sixth cervical vertebrae, C3–C6.

Insertion
Scalene tubercle on inner border of first rib.

Action
Acting together: flex neck. Raise first rib during active respiratory inhalation.
Individually: laterally flex and rotate neck.

Nerve
Ventral rami of cervical nerves, C5–C7.

Artery
Inferior thyroid artery of the thyrocervical trunk
(from subclavian artery).

Basic functional movement
Primarily a muscle of inspiration.

SCALENUS MEDIUS / SCALENUS POSTERIOR

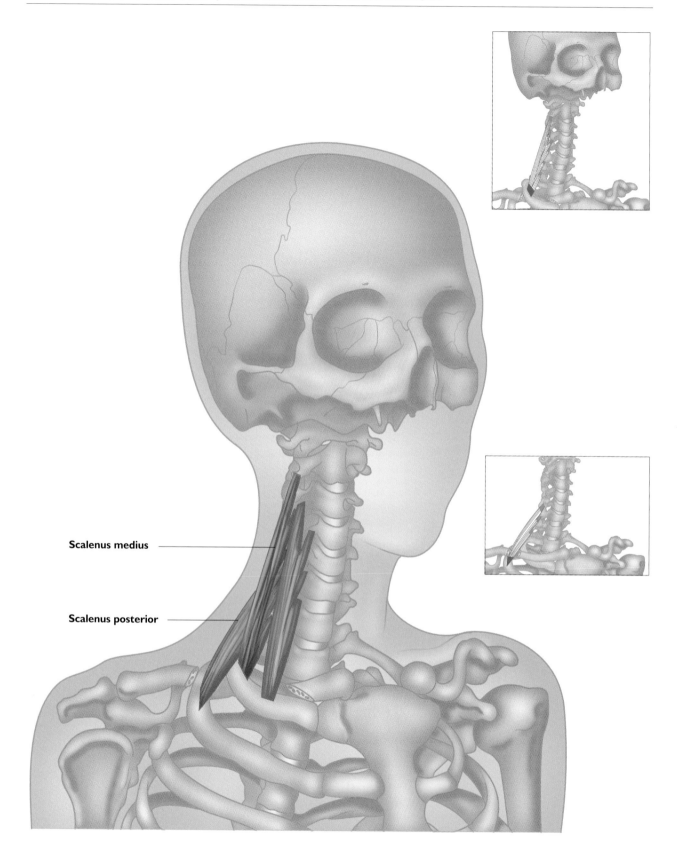

Scalenus medius

Scalenus posterior

Greek, *skalenos*, uneven; **Latin**, *medius*, middle.

Origin
Posterior tubercles of transverse processes of lower six cervical vertebrae, C2–C7.

Insertion
Upper surface of first rib, behind the groove for the subclavian artery.

Action
Acting together: flex neck. Raise first rib during active respiratory inhalation.
Individually: laterally flex and rotate neck.

Nerve
Ventral rami of cervical nerves, C3–C8.

Artery
Ascending cervical branch of the inferior thyroid artery
(from thyrocervical trunk of subclavian artery).

Basic functional movement
Primarily a muscle of inspiration.

Greek, *skalenos*, uneven; **Latin**, *posterior*, behind.

Origin
Posterior tubercles of transverse processes of lower two or three cervical vertebrae (C5–C7).

Insertion
Outer surface of second rib.

Action
Acting together: flex neck. Raise second rib during active respiratory inhalation.
Individually: laterally flex and rotate neck.

Nerve
Ventral rami of lower cervical nerves C7–C8.

Artery
Ascending cervical branch of the inferior thyroid artery
(from thyrocervical trunk of subclavian artery).

Basic functional movement
Primarily a muscle of inspiration.

STERNOCLEIDOMASTOIDEUS

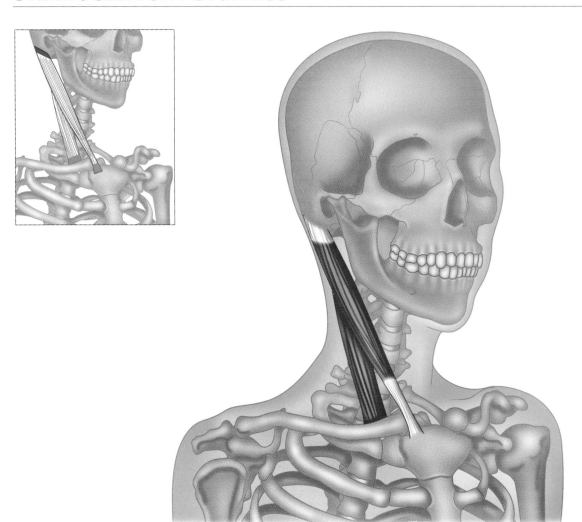

Greek, *sternon*, sternum; *kleidos*, key, clavicle; *mastoid*, breast-shaped, mastoid process.

This muscle is a long strap muscle with two heads. It is sometimes injured at birth, and may be partly replaced by fibrous tissue that contracts to produce a torticollis (wry neck).

Origin
Sternal head: anterior surface of manubrium of sternum.
Clavicular head: upper surface of medial third of clavicle.

Insertion
Outer surface of mastoid process of temporal bone. Lateral third of superior nuchal line of occipital bone.

Action
Contraction of both sides together: flexes neck and draws head forward, as in raising the head from a pillow. Raises sternum, and consequently the ribs, superiorly during deep inhalation.
Contraction of one side: tilts the head towards the same side. Rotates head to face the opposite side, (and also upwards as it does so).

Nerve
Accessory **X1** nerve; with sensory supply for proprioception from cervical nerves C2 and C3.

Artery
Sternocleidomastoid branches of the occipital artery, and superior thyroid arteries (from external carotid artery).

Basic functional movement
Examples: Turning head to look over your shoulder. Raising head from pillow.

Muscles of the Trunk

10

POSTVERTEBRAL MUSCLES

The postvertebral muscles are the deepest muscles of the back. They run longitudinally on the vertebral column. They are vital in maintaining posture and facilitating movements of the vertebral column. The fibres of the more superficial muscles of this group travel for a considerable distance between origin and insertion. The fibres of the deepest muscles stretch only between one vertebra and the next.

The erector spinae, also called sacrospinalis, comprises three sets of the postvertebral muscles organised in parallel columns. From lateral to medial, they are: *iliocostalis, longissimus* and *spinalis*.

Iliocostalis lumborum

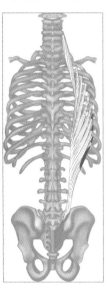

Iliocostalis cervicis

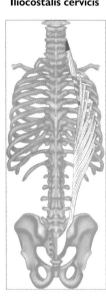

Longissimus cervicis

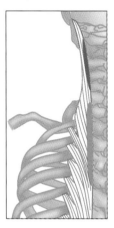

Spinalis thoracis

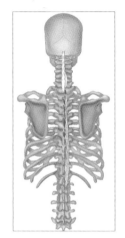

Splenius capitis

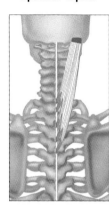

Iliocostalis thoracis

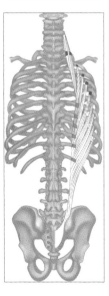

Longissimus thoracis

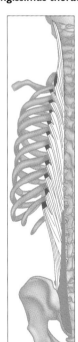

Longissimus capitis

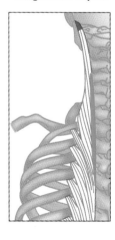

Spinalis cervicis

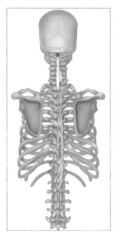

Splenius cervicis

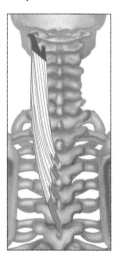

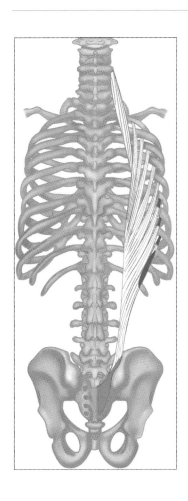

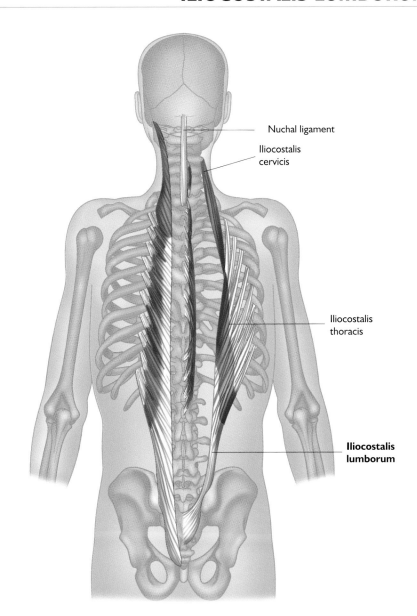

Nuchal ligament

Iliocostalis
cervicis

Iliocostalis
thoracis

**Iliocostalis
lumborum**

Latin, *iliocostalis*, from ilium to rib; *lumbar*, loin.

Iliocostalis is the most lateral part of the erector spinae. It may be subdivided into lumborum, thoracis and cervicis portions. As a whole, the iliocostalis is enervated via the dorsal rami of spinal nerves C4–S5.

Origin
Lateral and medial sacral crests. Medial part of iliac crests.

Insertion
Angles of lower six ribs.

Action
Extends and laterally flexes vertebral column. Helps maintain correct curvature of spine in the erect and sitting positions. Steadies the vertebral column on the pelvis during walking.

Nerve
Dorsal rami of lumbar nerves.

Artery
Lumbar arteries
(from abdominal aorta).
Subcostal arteries
(from thoracic aorta).

Basic functional movement
Keeps back straight (with correct curvatures).

ILIOCOSTALIS THORACIS / ILIOCOSTALIS CERVICIS

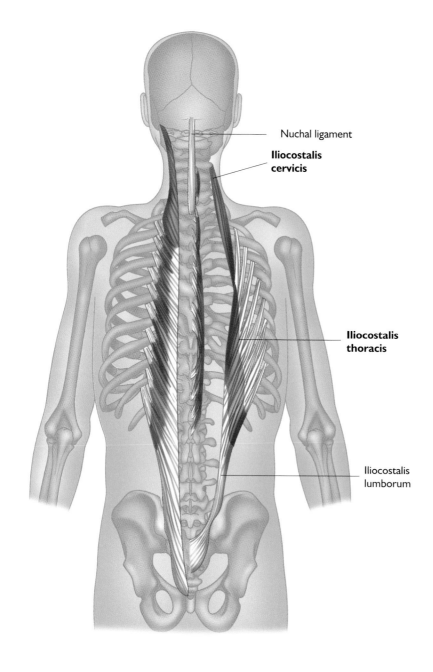

Nuchal ligament

Iliocostalis cervicis

Iliocostalis thoracis

Iliocostalis lumborum

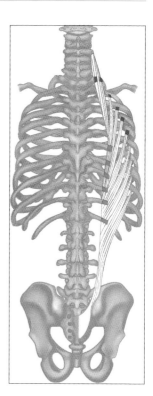

Iliocostalis thoracis

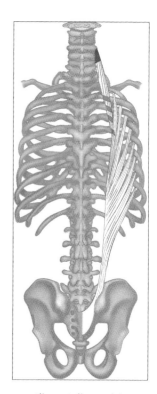

Iliocostalis cervicis

Latin, *iliocostalis*, from ilium to rib; *thoracicus*, pertaining to the chest.

Origin
Angles of lower six ribs, medial to iliocostalis lumborum.

Insertion
Angles of upper six ribs and transverse process of seventh cervical vertebra (C7).

Action
Extends and laterally flexes vertebral column. Helps maintain correct curvature of spine in the erect and sitting positions. Rotates ribs for forceful inhalation.

Nerve
Dorsal rami of thoracic (intercostal) nerves.

Artery
Posterior intercostal, and subcostal arteries
(from thoracic aorta).

Basic functional movement
Keeps back straight (with correct curvatures).

Latin, *iliocostalis*, from ilium to rib; *cervix*, neck.

Origin
Angles of third to sixth ribs.

Insertion
Transverse processes of fourth, fifth, and sixth cervical vertebrae (C4–C6).

Action
Extends and laterally flexes vertebral column. Helps maintain correct curvature of spine in the erect and sitting positions.

Nerve
Dorsal rami of cervical nerves.

Artery
Deep cervical artery of costocervical trunk
(from subclavian artery).

Basic functional movement
Keeps back straight (with correct curvatures).

LONGISSIMUS THORACIS / LONGISSIMUS CERVICIS

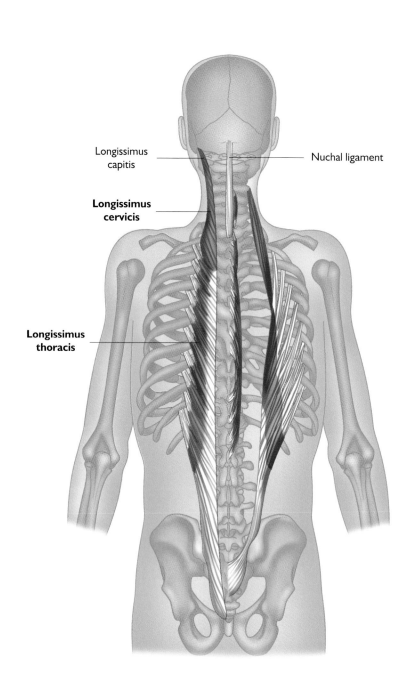

Longissimus
capitis

Longissimus
cervicis

Longissimus
thoracis

Nuchal ligament

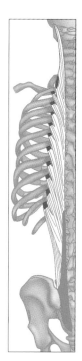

Longissimus
thoracis

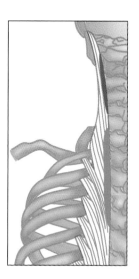

Longissimus cervicis

Latin, *longissimus*, longest; *thoracicus*, pertaining to the chest.

Longissimus is the intermediate part of the erector spinae. It may be subdivided into thoracis, cervicis and capitis portions. As a whole, the longissimus is enervated via the dorsal rami of spinal nerves C1–S1.

Origin
Lateral and medial sacral crests. Spinous processes and supraspinal ligament of all lumbar (L1–L5) and eleventh and twelfth thoracic vertebrae (T11–T12). Medial part of iliac crest.

Insertion
Transverse processes of all thoracic vertebrae (T1–T12). Area between tubercles and angles of lower nine or ten ribs.

Action
Extends and laterally flexes vertebral column. Helps maintain correct curvature of spine in the erect and sitting positions. Rotates ribs for forceful inhalation. Steadies the vertebral column on the pelvis during walking.

Nerve
Dorsal rami of spinal nerves.

Artery
Supplied segmentally by posterior intercostal and subcostal arteries
(from thoracic aorta).
Lumbar arteries
(from abdominal aorta).

Basic functional movement
Keeps back straight (with correct curvatures).

Latin, *longissimus*, longest; *cervix*, neck.

Longissimus is the intermediate part of the erector spinae. It may be subdivided into thoracis, cervicis and capitis portions. As a whole, the longissimus is enervated via the dorsal rami of spinal nerves C1–S1.

Origin
Transverse processes of upper four or five thoracic vertebrae (T1–T5).

Insertion
Transverse processes of second to sixth cervical vertebrae (C2–C6).

Action
Extends and laterally flexes upper vertebral column. Helps maintain correct curvature of thoracic and cervical spine in the erect and sitting positions.

Nerve
Dorsal rami of spinal nerves.

Artery
Supplied segmentally by deep cervical artery of costocervical trunk
(from subclavian artery).
Posterior intercostal arteries and subcostal arteries
(from thoracic aorta).

Basic functional movement
Keeps upper back and neck straight (with correct curvatures).

LONGISSIMUS CAPITIS

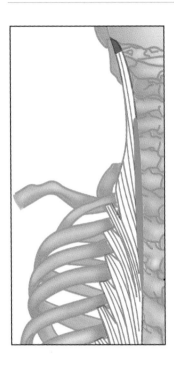

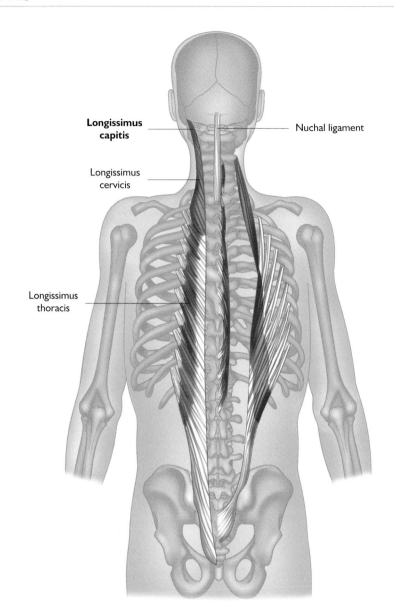

Longissimus
capitis

Nuchal ligament

Longissimus
cervicis

Longissimus
thoracis

Latin, *longissimus*, longest; *capitis*, of the head.

Longissimus is the intermediate part of the erector spinae. It may be subdivided into thoracis, cervicis and capitis portions. As a whole, the longissimus is enervated via the dorsal rami of spinal nerves C1–S1.

Origin
Transverse processes of upper five thoracic vertebrae (T1–T5). Articular processes of lower three cervical vertebrae (C5–C7).

Insertion
Posterior part of mastoid process of temporal bone.

Action
Extends and rotates head. Helps maintain correct curvature of thoracic and cervical spine in the erect and sitting positions.

Nerve
Dorsal rami of middle and lower cervical nerves.

Artery
Supplied segmentally by deep cervical artery of costocervical trunk
(from subclavian artery).
Posterior intercostal arteries and subcostal arteries
(from thoracic aorta).

Basic functional movement
Keeps upper back straight (with correct curvatures).

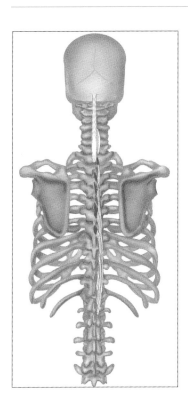

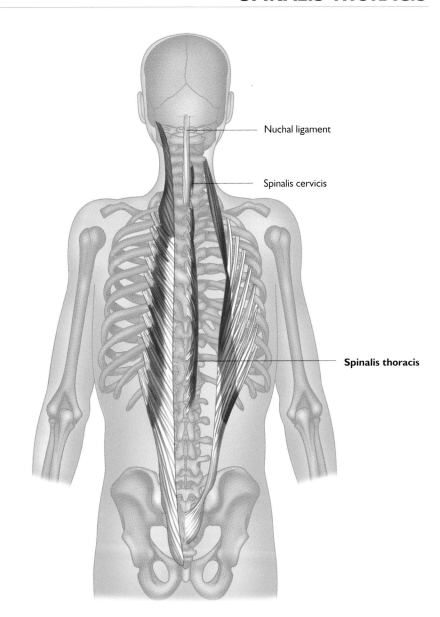

Nuchal ligament

Spinalis cervicis

Spinalis thoracis

Latin, *spinalis*, spinal; *thoracicus*, pertaining to the chest.

The spinalis is the most medial part of the erector spinae. It may be subdivided into thoracis, cervicis and capitis portions. As a whole, the spinalis is enervated via the dorsal rami of spinal nerves C2–L3.

Origin
Spinous processes of lower two thoracic (T11–T12) and upper two lumbar (L1–L2) vertebrae.

Insertion
Spinous processes of upper eight thoracic vertebrae (T1–T8).

Action
Extends vertebral column. Helps maintain correct curvature of spine in the erect and sitting positions.

Nerve
Dorsal rami of spinal nerves.

Artery
Supplied segmentally by posterior intercostal and subcostal arteries
(from thoracic aorta).
Lumbar arteries
(from abdominal aorta).

Basic functional movement
Keeps back straight (with correct curvatures).

SPINALIS CERVICIS

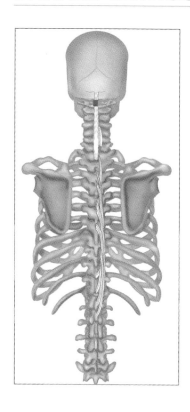

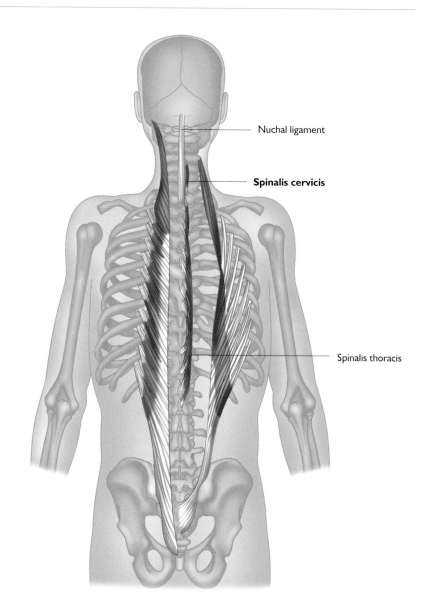

Nuchal ligament

Spinalis cervicis

Spinalis thoracis

Latin, *spinalis*, spinal; *cervix*, neck.

The spinalis is the most medial part of the erector spinae. It may be subdivided into thoracis, cervicis and capitis portions. As a whole, the spinalis is enervated via the dorsal rami of spinal nerves C2–L3.

Origin
Ligamentum nuchae. Spinous process of seventh cervical vertebra (C7).

Insertion
Spinous process of axis.

Action
Extends vertebral column. Helps maintain correct curvature of cervical spine in the erect and sitting positions.

Nerve
Dorsal rami of spinal nerves.

Artery
Supplied segmentally by the deep cervical artery of costocervical trunk
(from subclavian artery).

Basic functional movement
Keeps neck straight (with correct curvatures).

SPINALIS CAPITIS
(Medial part of Semispinalis Capitis)

Latin, *spinalis*, spinal; *capitis*, of the head.

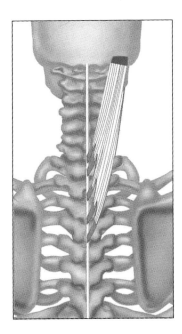

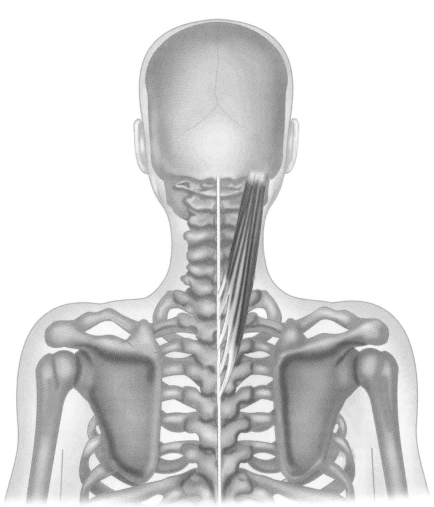

Greek, *splenion*, bandage; **Latin**, *capitis*, of the head.

Origin
Lower part of ligamentum nuchae. Spinous processes of the seventh cervical vertebra (C7) and upper three or four thoracic vertebrae (T1–T4).

Insertion
Posterior aspect of mastoid process of temporal bone. Lateral part of superior nuchal line, deep to the attachment of the sternocleidomastoideus.

Action
Acting together: extend the head and neck.
Individually: laterally flexes neck. Rotates the face to the same side as contracting muscle.

Nerve
Dorsal rami of middle and lower cervical nerves.

Artery
Supplied segmentally by deep cervical artery of costocervical trunk
(from subclavian artery).
Posterior intercostal arteries and the subcostal arteries
(from thoracic aorta).

Basic functional movement
Example: Looking up, or turning the head to look behind.

SPLENIUS CERVICIS

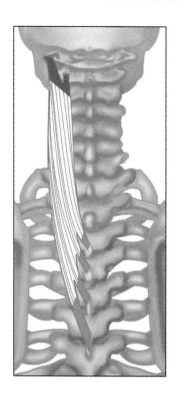

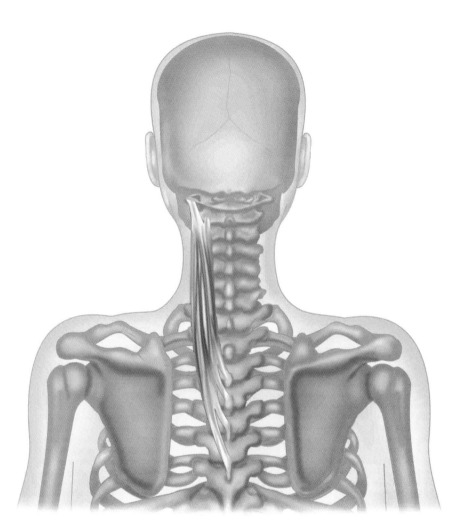

Greek, *splenion*, bandage; **Latin**, *cervix*, neck.

Origin
Spinous processes of the third to sixth thoracic vertebrae (T3–T6).

Insertion
Posterior tubercles of transverse processes of the upper two or three cervical vertebrae (C1–C3).

Action
Acting together: extend the head and neck.
Individually: laterally flexes neck. Rotates the face to the same side as contracting muscle.

Nerve
Dorsal rami of middle and lower cervical nerves.

Artery
Supplied segmentally by deep cervical artery of costocervical trunk
(from subclavian artery).
Posterior intercostal arteries and the subcostal arteries
(from thoracic aorta).

Basic functional movement
Example: Looking up, or turning the head to look behind.

The transversospinalis muscles are a composite of three small muscle groups situated deep to erector spinae. However, unlike erector spinae, each group lies successively deeper from the surface rather than side by side. The muscle groups are, from more superficial to deep: semispinalis, multifidis, and rotatores. Their fibres generally extend upward and medially from transverse processes to higher spinous processes.

Semispinalis thoracis

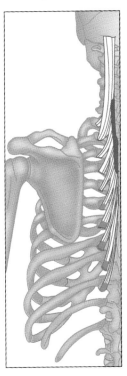

Semispinalis capitis

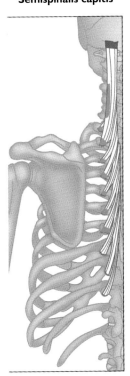

Semispinalis cervicis

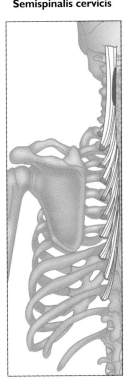

Rotatores

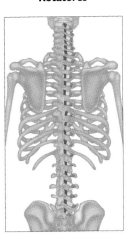

Multifidis

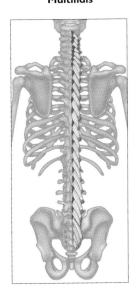

Interspinales

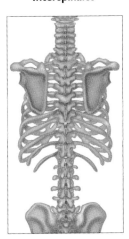

SEMISPINALIS THORACIS / SEMISPINALIS CERVICIS

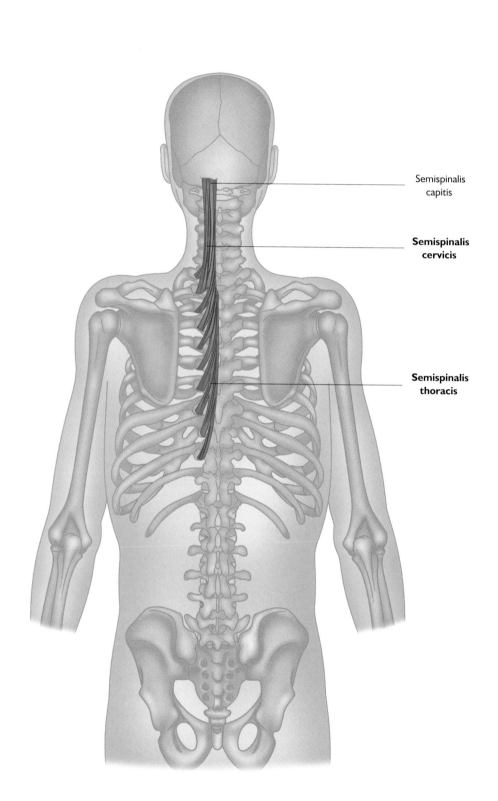

Semispinalis
capitis

**Semispinalis
cervicis**

**Semispinalis
thoracis**

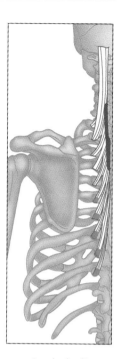

Semispinalis
thoracis

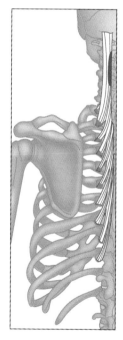

Semispinalis
cervicis

Latin, *semispinalis*, half spinal; *thoracicus*, pertaining to the chest.

Origin
Transverse processes of the sixth to tenth thoracic vertebrae (T6–T10).

Insertion
Spinous processes of the lower two cervical and upper four thoracic vertebrae (C6–T4).

Action
Extends thoracic and cervical parts of vertebral column. Assists rotation of thoracic and cervical vertebrae.

Nerve
Dorsal rami of thoracic and cervical spinal nerves.

Artery
Supplied segmentally by deep cervical artery of costocervical trunk
(from subclavian artery).
Posterior intercostal arteries and the subcostal arteries
(from thoracic aorta).

Basic functional movement
Example: Looking up, or turning the head to look behind.

Latin, *semispinalis*, half spinal; *cervix*, neck.

Origin
Transverse processes of upper five or six thoracic vertebrae (T1–T6).

Insertion
Spinous processes second to fifth cervical vertebrae (C2–C5).

Action
Extends thoracic and cervical parts of vertebral column. Assists rotation of thoracic and cervical vertebrae.

Nerve
Dorsal rami of thoracic and cervical spinal nerves.

Artery
Supplied segmentally by deep cervical artery of costocervical trunk
(from subclavian artery).
Posterior intercostal arteries and the subcostal arteries
(from thoracic aorta).

Basic functional movement
Example: Looking up, or turning the head to look behind.

SEMISPINALIS CAPITIS

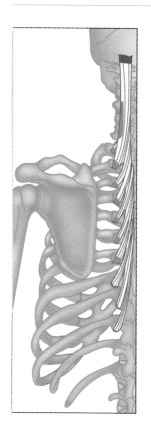

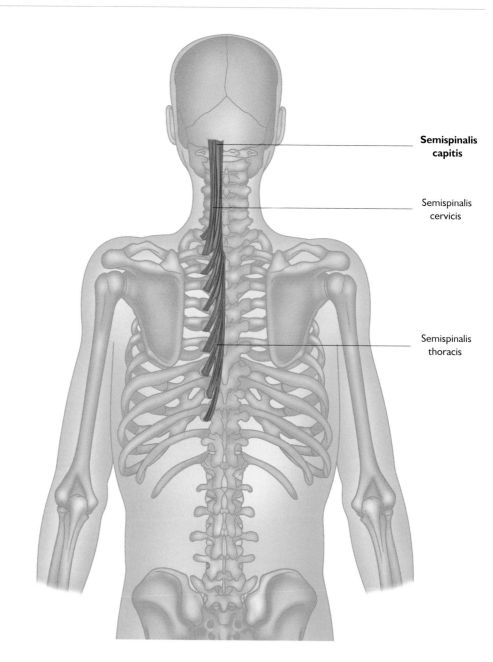

Semispinalis
capitis

Semispinalis
cervicis

Semispinalis
thoracis

Latin, *semispinalis*, half spinal; *capitis*, of the head.

Medial part is spinalis capitis.

Origin
Transverse processes of lower four cervical and upper six or seven thoracic vertebrae (C4–T7).

Insertion
Between superior and inferior nuchal lines of occipital bone.

Action
Most powerful extensor of the head. Assists in rotation of head.

Nerve
Dorsal rami of cervical nerves.

Artery
Supplied segmentally by deep cervical artery of costocervical trunk
(from subclavian artery).
Posterior intercostal arteries and the subcostal arteries
(from thoracic aorta).

Basic functional movement
Example: Looking up, or turning the head to look behind.

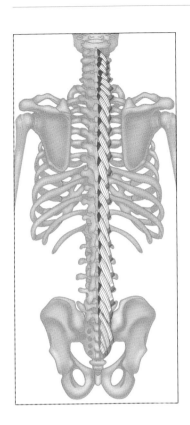

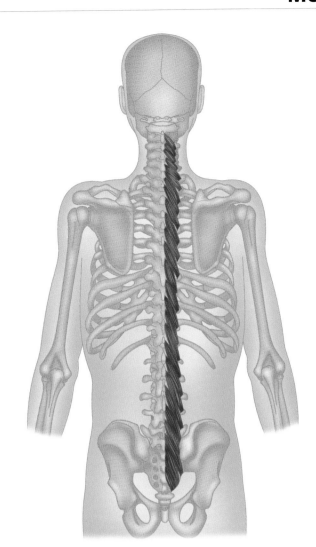

Latin, *multi*, many, much; *findere*, to split.

This muscle is the part of the transversospinalis group that lies in the furrow between the spines of the vertebrae and their transverse processes. It lies deep to semispinalis and erector spinae.

Origin
Posterior surface of sacrum, between the sacral foramina and posterior superior iliac spine. Mamillary processes (posterior borders of superior articular processes) of all lumbar vertebrae. Transverse processes of all thoracic vertebrae. Articular processes of lower four cervical vertebrae.

Insertion
Parts insert into spinous process two to four vertebrae superior to origin; overall including spinous processes of all the vertebrae from the fifth lumbar up to the axis (L5–C2).

Action
Protects vertebral joints from movements produced by the more powerful superficial prime movers. Extension, lateral flexion and rotation of vertebral column.

Artery
Supplied segmentally by deep cervical artery of costocervical trunk
(from subclavian artery).
Posterior intercostal arteries and the subcostal arteries
(from thoracic aorta).
Lumbar arteries
(from abdominal aorta).

Nerve
Dorsal rami of spinal nerves.

Basic functional movement
Helps maintain good posture and spinal stability during standing, sitting and all movements.

ROTATORES

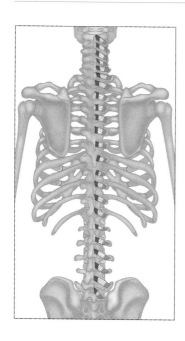

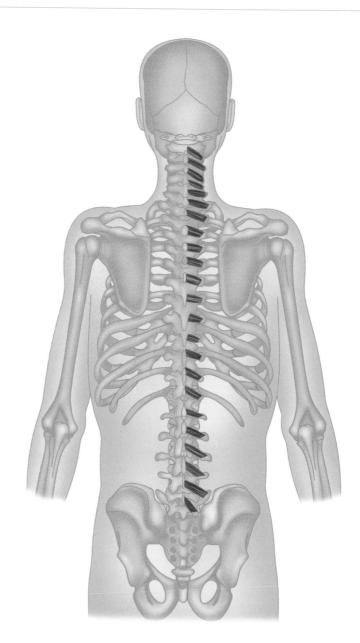

Latin, *rot*, wheel.

These small muscles are the deepest layer of the transversospinalis group.

Origin
Transverse process of each vertebra.

Insertion
Base of spinous process of adjoining vertebra above.

Action
Rotate and assist in extension of vertebral column.

Nerve
Dorsal rami of spinal nerves.

Artery
Supplied segmentally by deep cervical artery of costocervical trunk
(from subclavian artery).
Posterior intercostal arteries and the subcostal arteries
(from thoracic aorta).

Basic functional movement
Helps maintain good posture and spinal stability during standing, sitting and all movements.

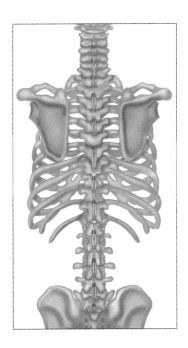

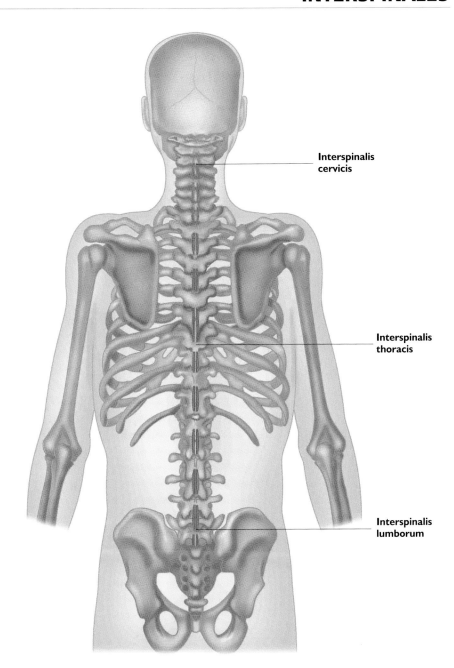

Interspinalis cervicis

Interspinalis thoracis

Interspinalis lumborum

Latin, *inter*, between; *spinalis*, spinal.

Short and insignificant muscles positioned either side of interspinous ligament. Interspinalis thoracis may be poorly developed or absent.

Origin / Insertion

Extend from one spinous process (origin) to the next one above (insertion), throughout the vertebral column. Most developed in cervical and lumbar regions. May be absent in the thoracic region.

Action

Act as extensile ligaments. Weakly extend vertebral column.

Nerve

Dorsal rami of spinal nerves.

Artery

Supplied segmentally by deep cervical artery of costocervical trunk
(from subclavian artery).
Posterior intercostal arteries and the subcostal arteries
(from thoracic aorta).

INTERTRANSVERSARII MUSCLES

Like the interspinales, the intertransversarii are also short and insignificant muscles. The cervical and thoracic regions, (the thoracic region is often poorly developed) encompass intertransversarii anteriores and intertransversarii posteriores, and the lumbar region encompasses intertransversarii laterales and intertransversarii mediales.

Intertransversarii anteriores

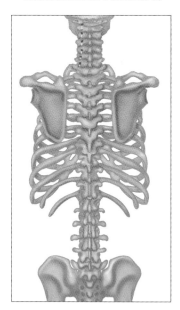

Intertransversarii posteriores

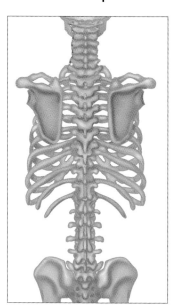

Intertransversarii laterales

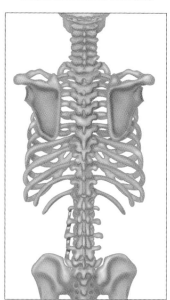

Intertransversarii mediales

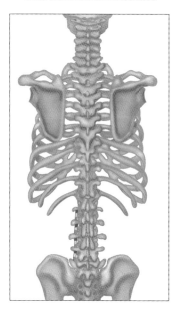

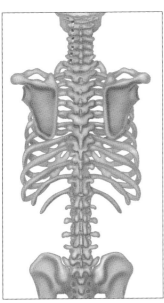

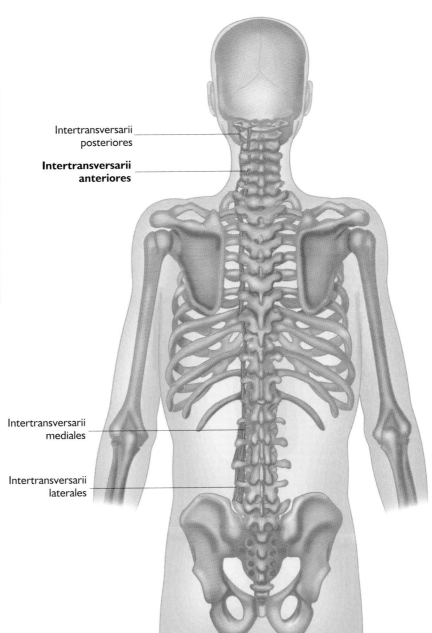

Intertransversarii posteriores

Intertransversarii anteriores

Intertransversarii mediales

Intertransversarii laterales

Latin, *inter*, between; *transverse*, across; *anterior*, before.

Origin
Anterior tubercle of transverse processes of vertebrae from first thoracic to axis (T1–C2).

Insertion
Anterior tubercle of next vertebra above.

Action
Slightly assists lateral flexion of cervical vertebrae. Act as extensile ligaments.

Nerve
Ventral rami of spinal nerves.

Artery
Supplied segmentally by the deep cervical artery of costocervical trunk
(from subclavian artery).

INTERTRANSVERSARII POSTERIORES / LATERALES

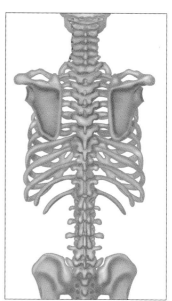

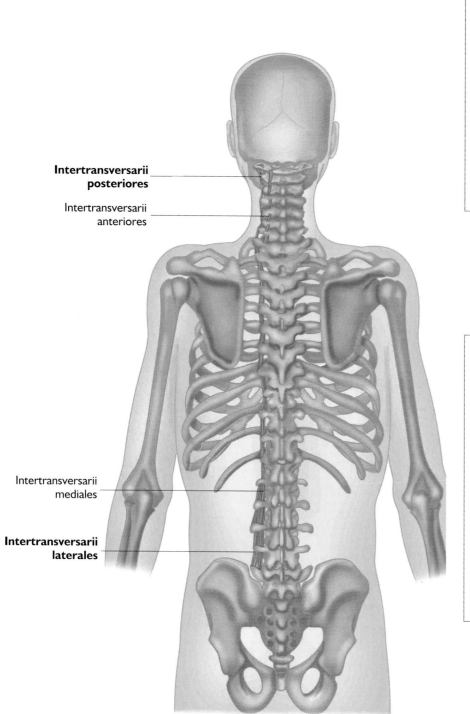

Intertransversarii posteriores

Intertransversarii anteriores

Intertransversarii mediales

Intertransversarii laterales

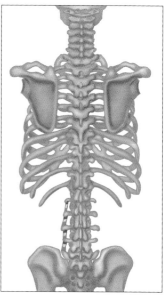

Latin, *inter*, between; *transverse*, across; *posterior*, behind.

Origin
Posterior tubercle of transverse processes of vertebrae from first thoracic to axis (T1–C2).
Transverse processes of first lumbar to eleventh thoracic vertebrae (L1–T11).

Insertion
Transverse process of next vertebra above (posterior tubercles in cervical region).

Action
In cervical region, slightly assists lateral flexion of cervical vertebrae. Act as extensile ligaments.

Nerve
Ventral rami of spinal nerves.

Artery
Supplied segmentally by the deep cervical artery of costocervical trunk
(from subclavian artery).
Posterior intercostal arteries and the subcostal arteries
(from thoracic aorta).

Latin, *inter*, between; *transverse*, across; *lateris*, side.

Origin
Transverse processes of lumbar vertebrae.

Insertion
Transverse process of next vertebra above.

Action
Slightly assists lateral flexion of lumbar vertebrae. Act as extensile ligaments.

Nerve
Ventral rami of spinal nerves.

Artery
Supplied segmentally by the lumbar arteries
(from abdominal aorta).

INTERTRANSVERSARII MEDIALES

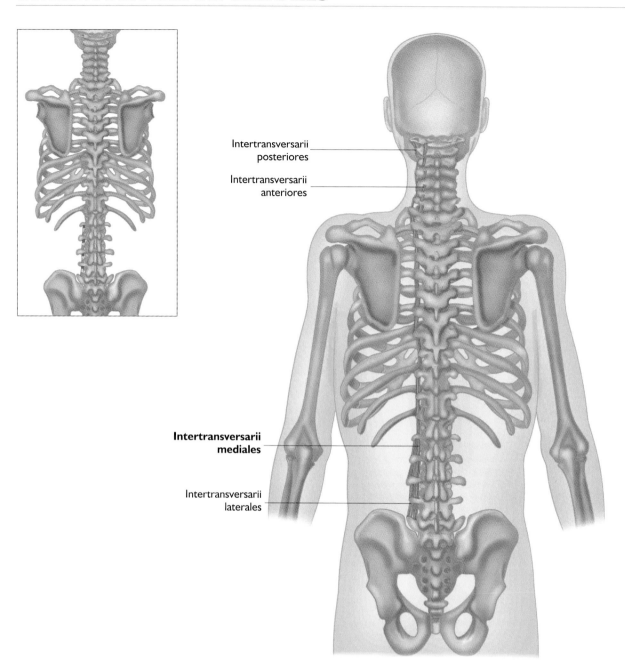

Intertransversarii
posteriores

Intertransversarii
anteriores

**Intertransversarii
mediales**

Intertransversarii
laterales

Latin, *inter*, between; *transverse*, across; *medial*, middle.

Origin
Mamillary process (posterior border of superior articular process) of each lumbar vertebra.

Insertion
Accessory process of next lumbar vertebra above.

Action
Slightly assists lateral flexion of lumbar vertebrae. Act as extensile ligaments.

Nerve
Dorsal rami of spinal nerves.

Artery
Supplied segmentally by the lumbar arteries (from abdominal aorta).

The suboccipital group of muscles lie deep in the neck anterior to the semispinalis capitis, longissimus capitis and splenius capitis. They enclose a triangular space known as the **suboccipital triangle**.

Rectus capitis posterior major

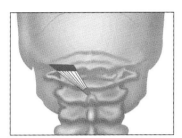

Obliquus capitis inferior

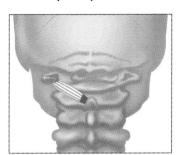

Rectus capitis posterior minor

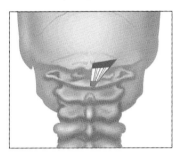

Obliquus capitis superior

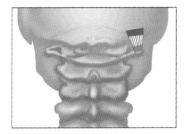

RECTUS CAPITIS POSTERIOR MAJOR

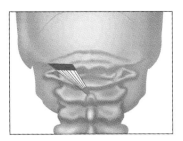

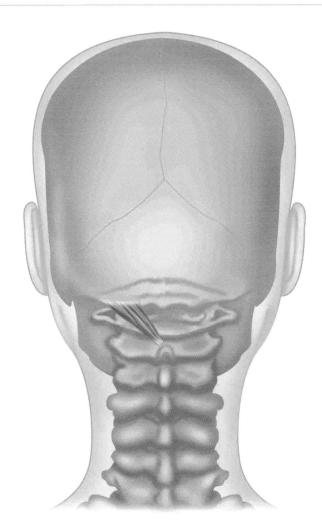

Latin, *rectum*, straight; *capitis*, of the head; *posterior*, behind; *major*, large.

Origin
Spinous process of axis.

Insertion
Below lateral portion of inferior nuchal line of occipital bone.

Action
Extends head. Rotates head to same side.

Nerve
Suboccipital nerve (dorsal ramus of first cervical nerve C1).

Artery
Occipital artery
(from external carotid artery).
Muscular branches of the vertebral artery
(from subclavian artery).

Basic functional movement
Helps smooth and stabilise the act of looking upwards and over the shoulder.

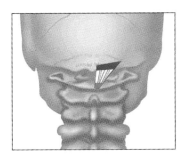

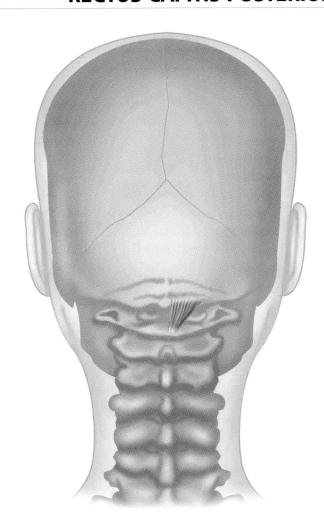

Latin, *rectum*, straight; *capitis*, of the head; *posterior*, behind; *minor*, small.

Origin
Posterior tubercle of atlas.

Insertion
Medial portion of inferior nuchal line of occipital bone.

Action
Extends head.

Nerve
Suboccipital nerve (dorsal ramus of first cervical nerve C1).

Artery
Occipital artery
(from external carotid artery).
Muscular branches of the vertebral artery
(from subclavian artery).

Basic functional movement
Helps smooth and stabilise the act of looking upwards.

OBLIQUUS CAPITIS INFERIOR

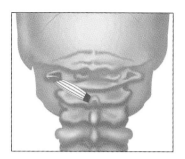

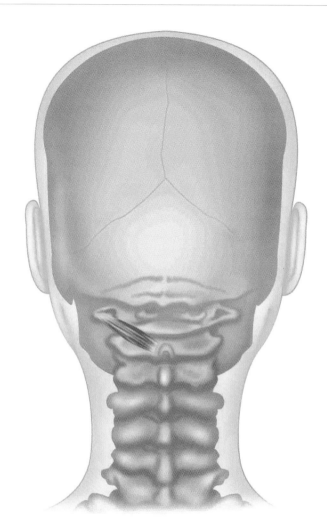

Latin, *obliquus*, inclined, slanting; *capitis*, of the head; *inferior*, below.

Origin
Spinous process of axis.

Insertion
Transverse process of atlas.

Action
Rotates atlas upon axis, thereby rotating head to the same side.

Nerve
Suboccipital nerve (dorsal ramus of first cervical nerve C1).

Artery
Occipital artery
(from external carotid artery).
Muscular branches of the vertebral artery
(from subclavian artery).

Basic functional movement
Gives stability to head when turning the head.

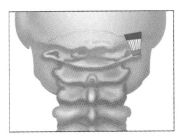

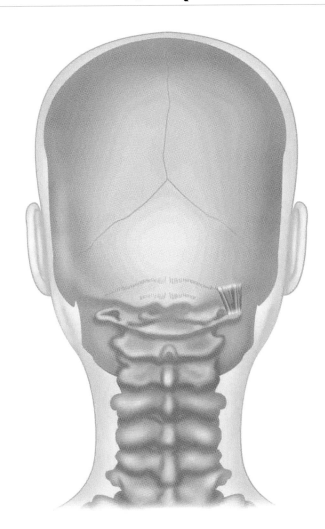

Latin, *obliquus,* inclined, slanting; *capitis,* of the head; *superior,* above.

Origin
Transverse process of atlas.

Insertion
Area between inferior and superior nuchal lines on occipital bone.

Action
Extends head.

Nerve
Suboccipital nerve (dorsal ramus of first cervical nerve C1).

Artery
Occipital artery
(from external carotid artery).
Muscular branches of the vertebral artery
(from subclavian artery).

Basic functional movement
Helps smooth and stabilise the act of looking upwards.

MUSCLES OF THE THORAX

The muscles in this section are small muscles, all primarily concerned with movements of the ribs.

Intercostales externi

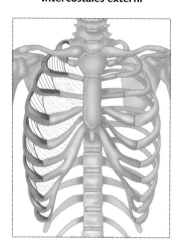

Intercostales interni

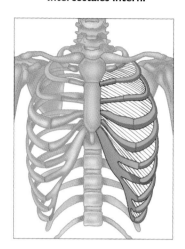

Subcostales

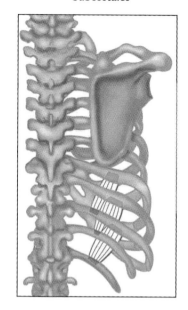

Levatores costarum

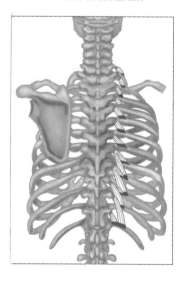

Diaphragm

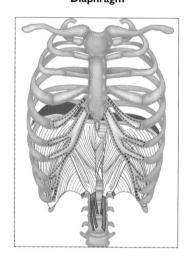

Transversus thoracis

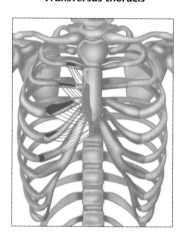

Serratus posterior superior

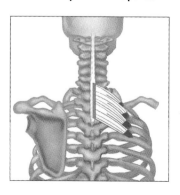

Serratus posterior inferior

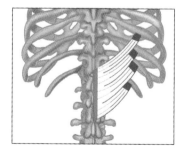

INTERCOSTALES EXTERNI (External Intercostal)

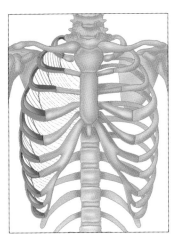

Anterior view.

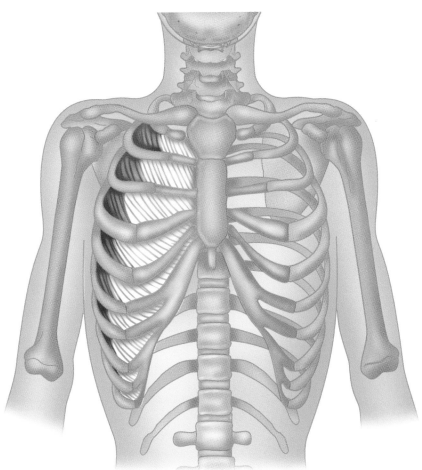

Latin, *inter*, between; *costal*, rib; *externi*, external.

The lower external intercostal muscles may blend with the fibres of external oblique, which overlap them, thus effectively forming one continuous sheet of muscle, with the external intercostal fibres seemingly stranded between the ribs. There are 11 external intercostals on each side of the ribcage.

Origin
Lower border of a rib.

Insertion
Upper border of rib below (fibres run obliquely forwards and downwards).

Action
Muscles contract to stabilize the ribcage during various movements of the trunk.
May elevate ribs during inspiration, thus increasing volume of thoracic cavity (although this action is disputed).
Prevents the intercostal space from bulging out or sucking in during respiration.

Nerve
The corresponding intercostal nerves.

Artery
Intercostal arteries
(from costocervical trunk of subclavian artery, and thoracic aorta).

INTERCOSTALES INTERNI (Internal Intercostal)

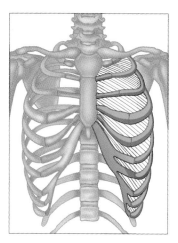

Anterior view.

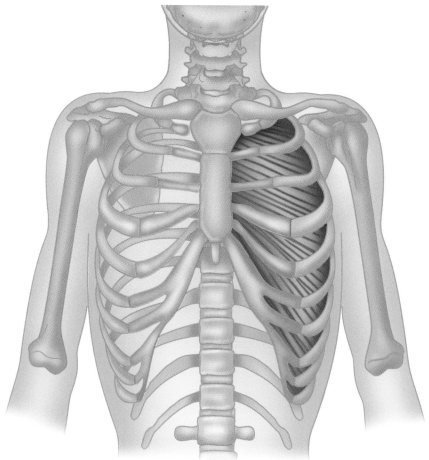

Latin, *inter*, between; *costal*, rib; *interni*, internal.

Internal intercostal fibres lie deep to, and run obliquely across, the external intercostals. There are 11 internal intercostals on each side of the ribcage.

Origin
Upper border of a rib and costal cartilage.

Insertion
Lower border of rib above (fibres run obliquely forwards and upwards towards the costal cartilage).

Action
Muscles contract to stabilize the ribcage during various movements of the trunk.
May draw adjacent ribs together during forced expiration, thus decreasing volume of thoracic cavity (although this action is disputed).
Prevents the intercostal space from bulging out or sucking in during respiration.

Nerve
The corresponding intercostal nerves.

Artery
Intercostal arteries
(from costocervical trunk of subclavian artery, and thoracic aorta).

These muscles are variable layers of fibres that run in the same direction as, but deep to, the internal intercostals. They are separated from internal intercostals by the intercostal nerve and vessels.

SUBCOSTALES

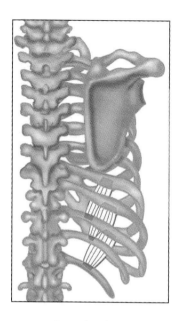

Posterior view.

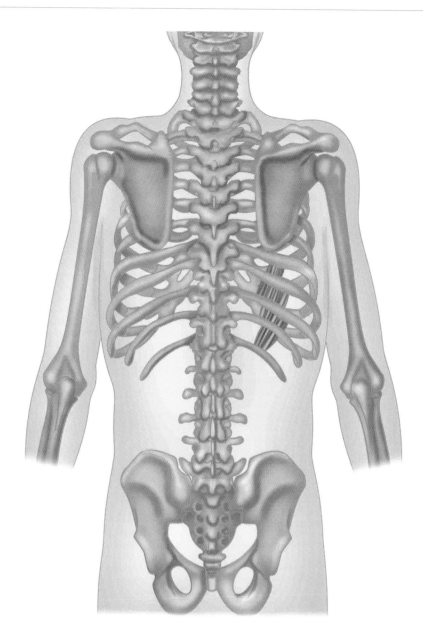

Latin, sub, under; *costal*, rib.

Positioned deep to the lower internal intercostals, the subcostales fibres run in the same direction as the innermost intercostal muscles and may be continuous with them. Subcostales, transversus thoracis and the innermost intercostal muscles make up the deepest intercostal muscle layer.

Origin
Inner surface of each lower rib near its angle.

Insertion
Fibres run obliquely medially into the inner surface of second or third rib below.

Action
Muscles contract to stabilize the ribcage during various movements of the trunk.
May draw adjacent ribs together during forced expiration, thus decreasing volume of thoracic cavity (although this action is disputed).

Nerve
The corresponding intercostal nerves.

Artery
Intercostal arteries
(from costocervical trunk of subclavian artery, and thoracic aorta).

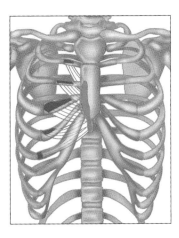

Anterior view.

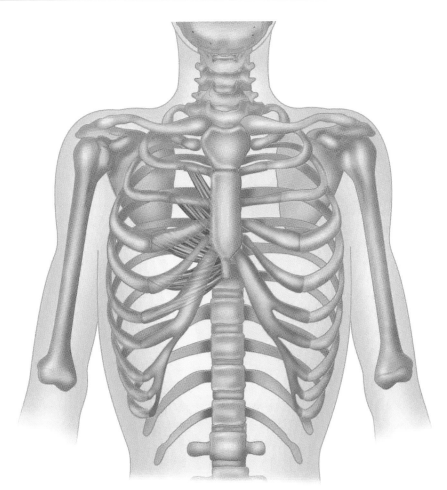

Latin, *transversus*, across, crosswise; *thoracicus*, pertaining to the chest.

Situated deep to internal intercostals.

Origin
Posterior surface of xiphoid process and body of sternum.

Insertion
Inner surfaces of costal cartilages of the second to sixth ribs.

Action
Draws costal cartilages downwards, contributing to forceful exhalation.

Nerve
The corresponding intercostal nerves.

Artery
Internal thoracic artery
(from subclavian artery).

Basic functional movement
Example: Blowing out a stubborn flame.

LEVATORES COSTARUM

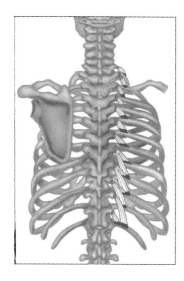

Posterior view.

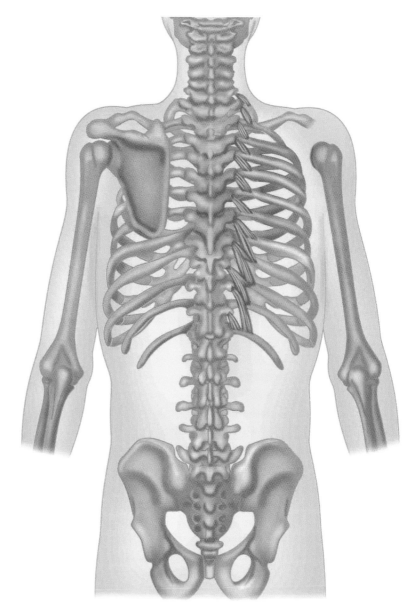

Latin, *levare,* to lift; *costarum,* of the rib.

Small, relatively insignificant muscles.

Origin
Transverse processes of seventh cervical to eleventh thoracic vertebra inclusive (C7–T11).

Insertion
Laterally downwards to external surface of the rib below, between the tubercle and the angle.

Action
Raises the ribs. May very slightly assist lateral flexion and rotation of vertebral column.

Nerve
Ventral rami of thoracic spinal nerves.

Artery
Deep cervical artery of costocervical trunk (from subclavian artery).
Intercostal arteries (from costocervical trunk of subclavian artery, and thoracic aorta).

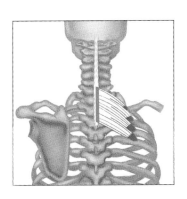

Posterior view.

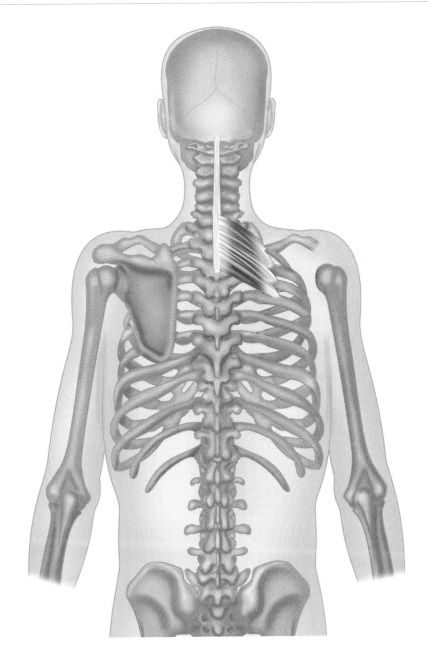

Latin, *serratus*, serrated; *posterior*, behind; *superior*, above.

Origin
Lower part of ligamentum nuchae. Spinous processes of seventh cervical (C7) and upper three or four thoracic vertebrae (T1–T4).

Insertion
Upper borders of second through to fifth ribs lateral to their angles.

Action
Raises the upper ribs (probably during forced inhalation).

Nerve
Intercostal nerves, T2, 3, 4

Artery
Intercostal arteries
(from costocervical trunk of subclavian artery, and thoracic aorta).

SERRATUS POSTERIOR INFERIOR

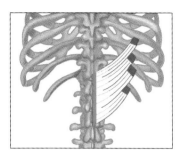

Posterior view.

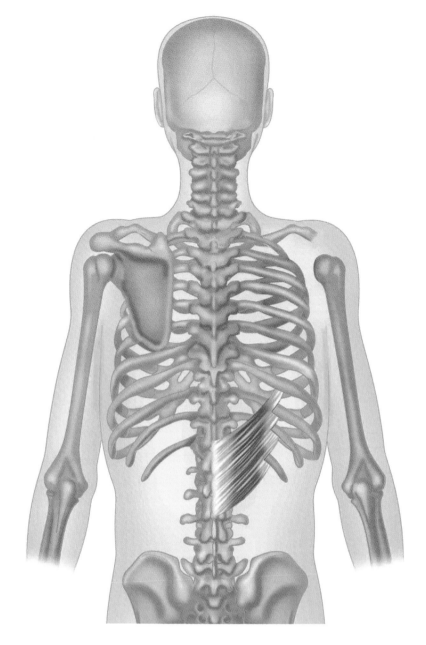

Latin, *serratus*, serrated; *posterior*, behind; *inferior*, below.

Origin
Thoracolumbar fascia, at its attachment to the spinous processes of the lower two thoracic (T11–T12) and upper two or three lumbar vertebrae (L1–L3).

Insertion
Lower borders of last four ribs.

Action
May help draw lower ribs downwards and backwards, resisting the pull of the diaphragm.

Nerve
Intercostal nerves, T9, 10, 11.

Artery
Intercostal arteries
(from thoracic aorta).

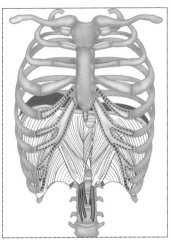

Origin on posterior
of costal cartilage.

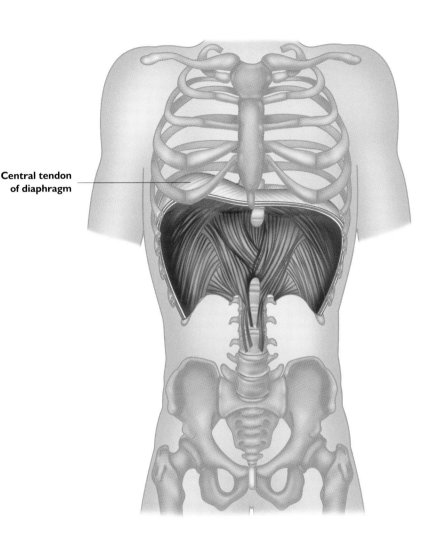

**Central tendon
of diaphragm**

Greek, partition, wall.

Origin

Sternal portion: back of xiphoid process.

Costal portion: inner surfaces of lower six ribs and their costal cartilages.

Lumbar portion: upper two or three lumbar vertebrae (L1–L3). Medial and lateral lumbocostal arches (also known as the medial and lateral arcuate ligaments).

Insertion

All fibres converge and attach onto a central tendon; i.e. this muscle inserts upon itself.

Action

Forms floor of thoracic cavity. Pulls central tendon downwards during inhalation, thereby increasing volume of thoracic cavity.

Nerve

Phrenic nerve (ventral rami), C3, **4**, 5.

Artery

Musculophrenic artery
via internal thoracic artery (from subclavian artery).

Superior phrenic artery
(from thoracic aorta).

Inferior phrenic artery
(from abdominal aorta).

Basic functional movement

Produces about 60% of your breathing capacity.

MUSCLES OF THE ANTERIOR ABDOMINAL WALL

The anterior abdominal wall has three layers of muscle, with fibres running in the same direction as the corresponding three layers of muscle in the thoracic wall. The deepest layer consists of the transversus abdominis, whose fibres run approximately horizontally. The middle layer comprises the internal oblique, whose fibres are crossed by the outermost layer known as the external oblique, forming a pattern of fibres resembling a St. Andrew's cross. Overlying these three layers is the rectus abdominis, which runs vertically, either side of the midline of the abdomen.

Obliquus externus abdominis

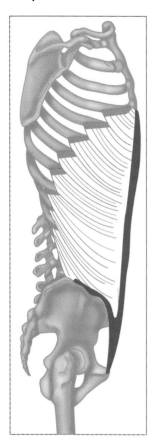

Obliquus internus abdominis

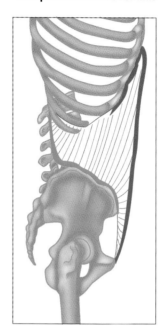

Transversus abdominis

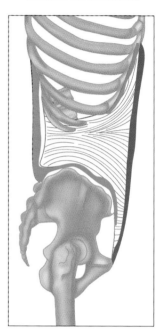

Cremaster

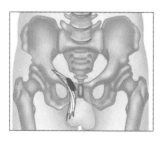

Rectus abdominis

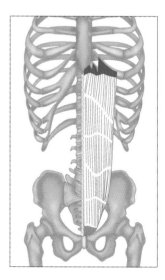

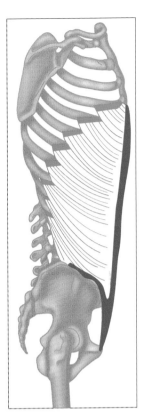

Lateral view.

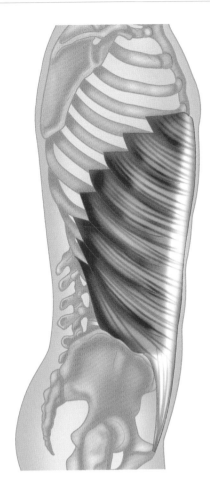

Latin, *obliquus*, inclined, slanting; *externus*, external; *abdominis*, belly / stomach.

The posterior fibres of the external oblique are usually overlapped by the latissimus dorsi, but in some cases there is a space between the two, known as the lumbar triangle, situated just above the iliac crest. The **lumbar triangle** is a weak point in the abdominal wall.

Origin
Anterior fibres: outer surfaces of ribs five through eight, interdigitating with serratus anterior.
Lateral fibres: outer surface of ninth rib, interdigitating with serratus anterior; and outer surfaces of tenth, eleventh and twelfth ribs, interdigitating with latissimus dorsi.

Insertion
Anterior fibres: into a broad, flat abdominal aponeurosis that terminates in the linea alba, a tendinous raphe extending from the xiphoid process.
Lateral fibres: as the inguinal ligament, into the anterior superior iliac spine and the pubic tubercle, and into the lip of the anterior one-half of the iliac crest.

Action
Compresses abdomen, helping to support the abdominal viscera against the pull of gravity. Contraction of one side alone bends the trunk laterally to that side and rotates it to the opposite side.

Nerve
Ventral rami of thoracic nerves, T5–T12.

Artery
Musculophrenic artery and superior epigastric artery (via internal thoracic artery from subclavian artery).
Intercostal arteries 7–11 and subcostal artery (from thoracic aorta).
Lumbar arteries (from abdominal aorta).
Superficial circumflex artery, superficial epigastric artery and the superficial external pudendal artery (from femoral artery).
Deep circumflex iliac artery and inferior epigastric artery (from external iliac artery).

Basic functional movement
Example: Digging with a shovel.

OBLIQUUS INTERNUS ABDOMINIS (Internal Oblique)

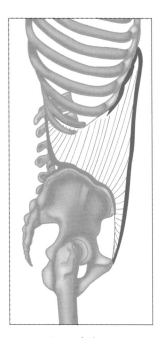

Lateral view.

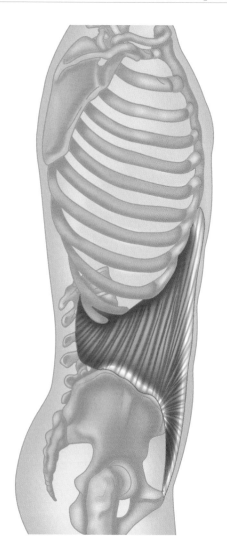

Latin, *obliquus*, inclined, slanting; *internus*, internal; *abdominis*, belly / stomach.

Origin
Iliac crest. Lateral two-thirds of inguinal ligament. Thoracolumbar fascia.

Insertion
Inferior borders of bottom three or four ribs. Linea alba via an abdominal aponeurosis. Crest of pubis (along with transversus abdominis).

Action
Compresses abdomen, helping to support the abdominal viscera against the pull of gravity. Contraction of one side alone laterally bends and rotates the trunk.

Nerve
Ventral rami of thoracic nerves, T7–T12, ilioinguinal and iliohypogastric nerves.

Artery
Musculophrenic artery and superior epigastric artery
(via internal thoracic artery from subclavian artery).
Intercostal arteries 7–11 and subcostal artery
(from thoracic aorta).
Lumbar arteries
(from abdominal aorta).
Superficial circumflex artery, superficial epigastric artery and the superficial external pudendal artery
(from femoral artery).
Deep circumflex iliac artery and inferior epigastric artery
(from external iliac artery).

Basic functional movement
Example: Raking.

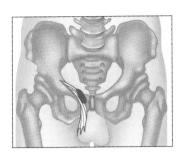

Anterior view.

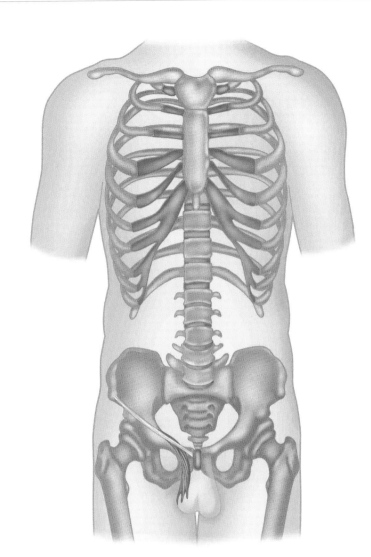

Greek, *kremasthai*, to suspend.

In males, the cremaster is usually well developed. In females, it is underdeveloped or absent. It forms a thin network of muscle fibres around the spermatic cord and testes (or around the distal portion of the round ligament of the uterus).

Origin
Inguinal ligament.

Insertion
Pubic tubercle. Crest of pubis. Sheath of rectus abdominis.

Action
Pulls testes up from the scrotum towards the body (mainly to regulate the temperature of the testes).

Nerve
Genital branch of genitofemoral nerve, L1, 2.

Artery
Cremasteric artery
which is a branch of the inferior epigastric artery (from external iliac artery).

TRANSVERSUS ABDOMINIS

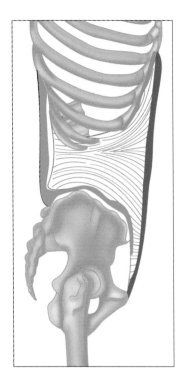

Lateral view.

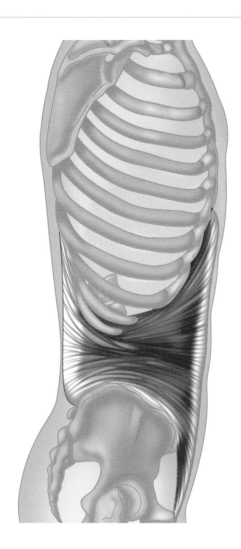

Latin, *transversus*, across, crosswise; *abdominis*, belly / stomach.

Origin
Anterior two thirds of iliac crest. Lateral third of inguinal ligament. Thoracolumbar fascia. Costal cartilages of lower six ribs. Fascia covering iliopsoas.

Insertion
Xiphoid process and linea alba via an abdominal aponeurosis, the lower fibres of which ultimately attach to the pubic crest and pecten pubis via the conjoint tendon.

Action
Compresses abdomen, helping to support the abdominal viscera against the pull of gravity.

Nerve
Ventral rami of thoracic nerves T7–T12, ilioinguinal and iliohypogastric nerves.

Artery
Musculophrenic artery and superior epigastric artery
via internal thoracic artery (from subclavian artery).
Intercostal arteries 7–11 and subcostal artery
(from thoracic aorta).
Lumbar arteries
(from abdominal aorta).
Superficial circumflex artery, superficial epigastric artery and the superficial external pudendal artery
(from femoral artery).
Deep circumflex iliac artery and inferior epigastric artery
(from external iliac artery).

Basic functional movement
Important during forced expiration, sneezing and coughing. Helps maintain good posture.

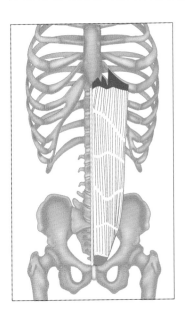

Anterior view.

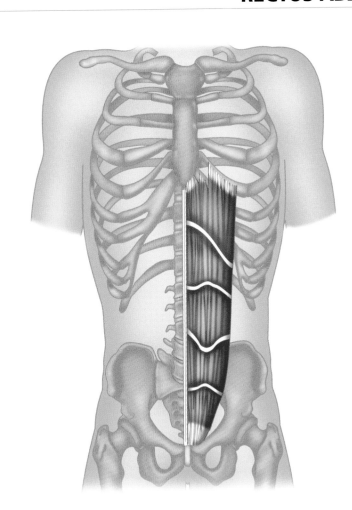

Latin, *rectum*, straight; *abdominis*, belly / stomach.

The rectus abdominis is divided by tendinous bands into three or four bellies, each sheathed in aponeurotic fibres from the lateral abdominal muscles. These fibres converge centrally to form the linea alba. Situated anterior to the lower part of rectus abdominis is a frequently absent muscle called *pyramidalis,* which arises from the pubic crest and inserts into the linea alba. It tenses the linea alba, for reasons unknown.

Origin
Pubic crest and symphysis pubis.

Insertion
Anterior surface of xiphoid process. Fifth, sixth and seventh costal cartilages.

Action
Flexes lumbar spine. Depresses ribcage. Stabilizes the pelvis during walking.

Nerve
Ventral rami of thoracic nerves, T5–12.

Artery
Superior epigastric artery
via internal thoracic artery (from subclavian artery).
Intercostal arteries and subcostal artery
(from thoracic aorta).
Inferior epigastric artery
(from external iliac artery).

Basic functional movement
Example: Initiating getting out of a low chair.

MUSCLES OF THE POSTERIOR ABDOMINAL WALL

The posterior abdominal wall comprises the *quadratus lumborum*, with the origin of *psoas major* positioned medial to it, covering the sides of the lumbar vertebral bodies and the anterior aspects of their transverse processes. The psoas major runs downwards to be joined by the *iliacus*, which lines the iliac fossa. Together, these muscles act as padding for various abdominal viscera, and leave the abdomen to become the main flexor of the hip joint.

Quadratus lumborum

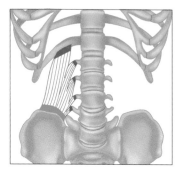

Psoas major

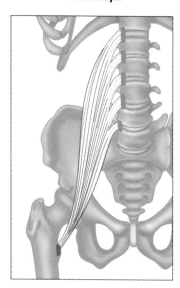

Iliacus

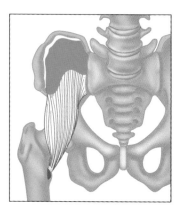

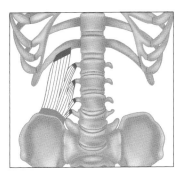

Anterior view.

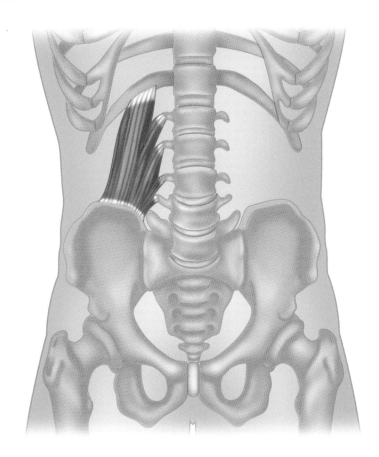

Latin, *quadratus*, squared; *lumbar*, loin.

Origin
Posterior part of iliac crest. Iliolumbar ligament.

Insertion
Medial part of lower border of twelfth rib. Transverse processes of upper four lumbar vertebrae (L1–L4).

Action
Laterally flexes vertebral column. Fixes the twelfth rib during deep respiration (e.g. helps stabilise the diaphragm for singers exercising voice control). Helps extend lumbar part of vertebral column, and gives it lateral stability.

Nerve
Ventral rami of the subcostal nerve and upper three or four lumbar nerves, T12, L**1**, **2**, **3**.

Artery
Subcostal artery
(from thoracic aorta).
Lumbar arteries
(from abdominal aorta).

Basic functional movement
Example: Bending sideways from sitting to pick up an object from the floor.

PSOAS MAJOR (Part of Iliopsoas)

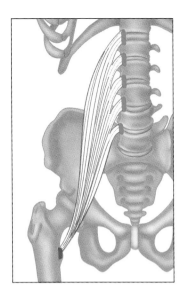

Anterior view.

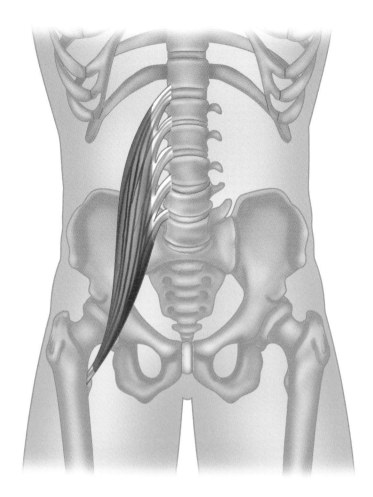

Greek, *psoas*, muscle of loin; *major*, large.

The psoas major and iliacus are considered part of the posterior abdominal wall due to their position and cushioning role for the abdominal viscera. However, based on their action of flexing the hip joint, it would also be relevant to place them in the section entitled Muscles of the Hip (*see* page 317–323). Note that some upper fibres of psoas major may insert by a long tendon into the iliopubic eminence to form the psoas minor, which has little function and is absent in about 40% of people.

Bilateral contracture of this muscle will increase lumbar lordosis.

Origin
Bases of transverse processes of all lumbar vertebrae (L1–L5). Bodies of twelfth thoracic and all lumbar vertebrae (T12–L5). Intervertebral discs above each lumbar vertebra.

Insertion
Lesser trochanter of femur.

Action
Main flexor of hip joint, in conjunction with iliacus (flexes and laterally rotates thigh, as in kicking a football). Acting from its insertion, flexes the trunk, as in sitting up from the supine position.

Nerve
Ventral rami of lumbar nerves, L1, **2**, **3**, 4 (psoas minor innervated from L**1**, **2**).

Artery
Subcostal artery
(from thoracic aorta).
Lumbar arteries
(from abdominal aorta).

Basic functional movement
Example: Going up a step or walking up an incline.

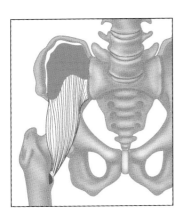

Anterior view.

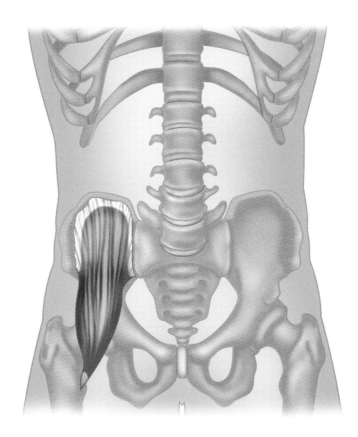

Latin, pertaining to the loin.

Origin
Superior two-thirds of iliac fossa. Internal lip of iliac crest. Ala of sacrum and anterior ligaments of the lumbosacral and sacroiliac joints.

Insertion
Lateral side of tendon of psoas major, continuing into lesser trochanter of femur.

Action
Main flexor of hip joint, in conjunction with psoas major (flexes and laterally rotates thigh, as in kicking a football. Brings leg forward in walking or running). Acting from its insertion, flexes the trunk, as in sitting up from the supine position.

Nerve
Femoral nerve, L (1), **2, 3**, 4.

Artery
Iliolumbar branch of the internal iliac artery
via common iliac artery (from abdominal aorta).

Basic functional movement
Example: Going up a step or walking up an incline.

Muscles of the Shoulder and Arm

11

Muscles Attaching
the Upper Limb
to the Trunk

Muscles of the
Shoulder Joint

Muscles of the Arm

MUSCLES ATTACHING THE UPPER LIMB TO THE TRUNK

Mobility of the upper limb is mainly dependent on three joints: the sternoclavicular, acromioclavicular, and shoulder joints. Muscles in this area can be categorised according to: 1) Muscles that run between the trunk and the scapula, which act upon the shoulder girdle and *not* on the shoulder joint: i.e. *trapezius, levator scapulae, rhomboids, serratus anterior, pectoralis minor* and *subclavius*. 2) Muscles that run between the trunk and the humerus, which act upon the shoulder joint *and* the shoulder girdle: i.e. *latissimus dorsi* and *pectoralis major*. 3) Muscles that run between the scapula and humerus, which act exclusively upon the shoulder joint: i.e. *deltoid, supraspinatus, infraspinatus, teres minor, subscapularis, teres major*, and *coracobrachialis*.

Trapezius

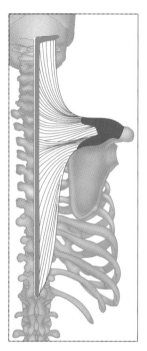

Rhomboideus minor

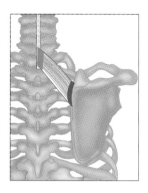

Rhomboideus major

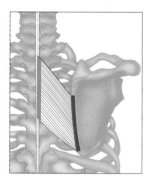

Latissimus dorsi

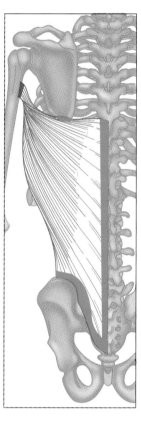

Pectoralis major

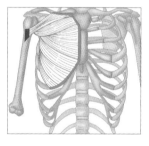

Pectoralis minor

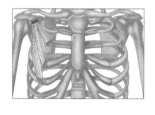

Levator scapulae

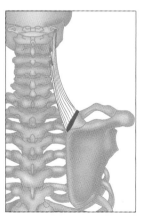

Serratus anterior

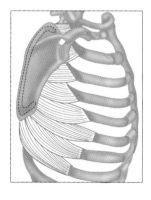

Subclavius

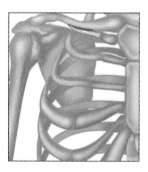

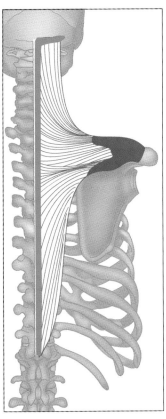

Posterior view.

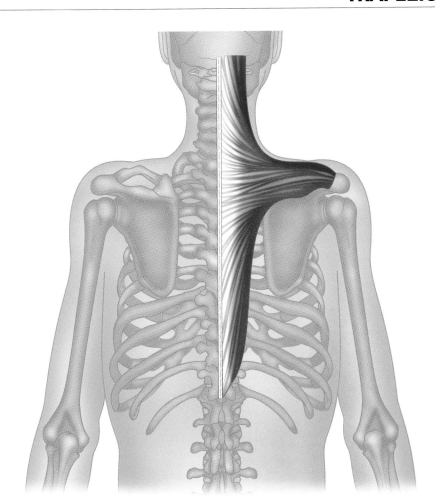

Latin, *trapezoides*, table shaped.

The left and right trapezius viewed as a whole, create a trapezium in shape, thus giving this muscle its name.

Origin
Medial third of superior nuchal line of occipital bone. External occipital protuberance. Ligamentum nuchae. Spinous processes and supraspinous ligaments of seventh cervical (C7) and all thoracic vertebrae (T1–T12).

Insertion
Posterior border of lateral third of clavicle. Medial border of acromion. Upper border of the crest of the spine of scapula, and the tubercle on this crest.

Action
Upper fibres: pull the shoulder girdle up (elevation). Helps prevent depression of the shoulder girdle when a weight is carried on the shoulder or in the hand.

Middle fibres: retract (adduct) scapula.
Lower fibres: depress scapula, particularly against resistance, as when using the hands to get up from a chair.
Upper and lower fibres together: rotate scapula, as in elevating the arm above the head.

Nerve
Motor supply: Accessory **X1** nerve.
Sensory supply (proprioception): Ventral ramus of cervical nerves, C2, **3**, **4**.

Artery
Transverse cervical artery
(from subclavian artery).

Basic functional movement
Example (upper and lower fibres working together): Painting a ceiling.

LEVATOR SCAPULAE

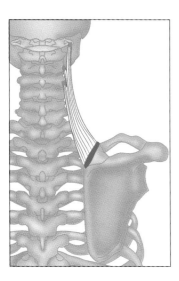

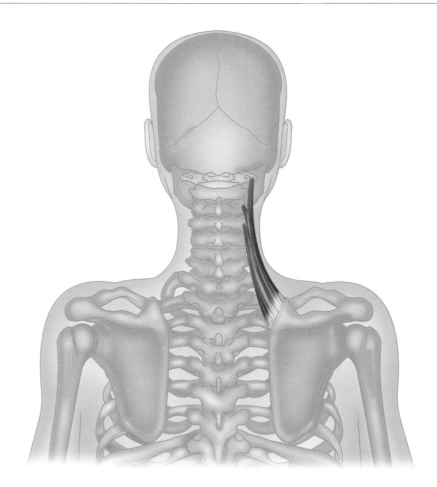

Latin, *levare*, to lift; *scapulae*, shoulder, blade(s).

Levator scapulae is deep to sternocleidomastoideus and trapezius. It is named after its action of elevating the scapula.

Origin
Posterior tubercles of the transverse processes of the first three or four cervical vertebrae (C1–C4).

Insertion
Medial (vertebral) border of the scapula between the superior angle and the spine of scapula.

Action
Elevates scapula. Helps retract scapula. Helps bend neck laterally.

Nerve
Dorsal scapular nerve, C**4**, **5** and cervical nerves C**3**, **4**.

Artery
Dorsal scapular artery
via deep branch of transverse cervical artery (from subclavian artery).

Basic functional movement
Example: Carrying a heavy bag.

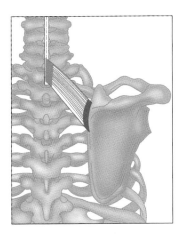

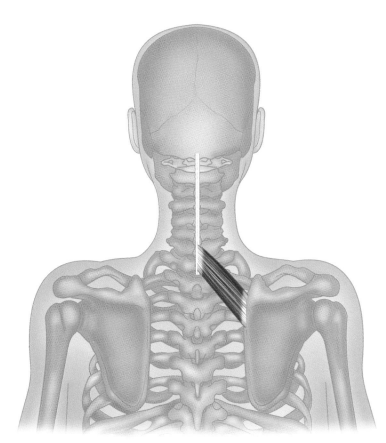

Greek, *rhomb*, a parallelogram with oblique angles and only the opposite sides equal; *minor*, small.

So named because of its shape.

Origin
Spinous processes and supraspinous ligaments of the seventh cervical and first thoracic vertebrae. Lower part of ligamentum nuchae.

Insertion
Medial (vertebral) border of scapula at the level of the spine of scapula.

Action
Retracts (adducts) scapula. Stabilizes scapula. Slightly elevates medial border of the scapula causing downward rotation (therefore depressing the lateral angle). Slightly assists in outer range of adduction of arm (i.e. from arm overhead to arm at shoulder level).

Nerve
Dorsal scapular nerve, C**4**, **5**.

Artery
Dorsal scapular artery
 via deep branch of transverse cervical artery (from subclavian artery).

Basic functional movement
Example: Pulling something towards you, such as opening a drawer.

RHOMBOIDEUS MAJOR

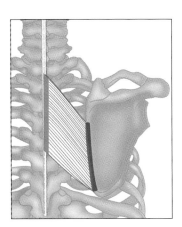

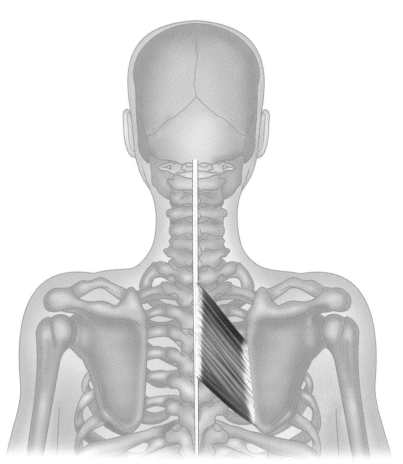

Greek, *rhomb*, a parallelogram with oblique angles and only the opposite sides equal; *major*, large.

The rhomboideus major runs parallel to, and is often continuous with, the rhomboideus minor. It is so named because of its shape.

Origin
Spinous processes and supraspinous ligaments of second to fifth thoracic vertebrae (T2–T5).

Insertion
Medial border of the scapula, between the spine of scapula and the inferior angle.

Action
Retracts (adducts) scapula. Stabilizes scapula. Slightly elevates medial border of the scapula causing downward rotation. Slightly assists in outer range of adduction of arm (i.e. from arm overhead to arm at shoulder level).

Nerve
Dorsal scapular nerve, C**4**, **5**.

Artery
Dorsal scapular artery
via deep branch of transverse cervical artery (from subclavian artery).

Basic functional movement
Example: Pulling something towards you, such as opening a drawer.

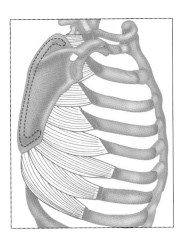

Insertion on anterior of scapula.
Lateral view.

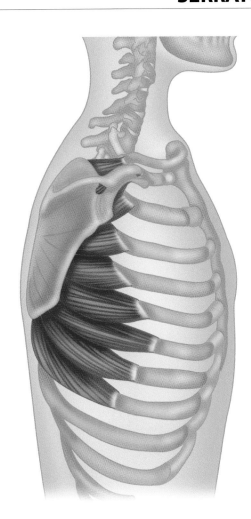

Latin, *serratus*, serrated; *anterior*, before.

The serratus anterior forms the medial wall of the axilla, along with the upper five ribs. It is a large muscle composed of a series of finger-like slips. The lower slips interdigitate with the origin of the external oblique.

Origin
Outer surfaces and superior borders of upper eight or nine ribs, and the fascia covering their intercostal spaces.

Insertion
Anterior (costal) surface of the medial border of scapula and inferior angle of scapula.

Action
Rotates scapula for abduction and flexion of arm. Protracts scapula (pulls it forward on the chest wall and holds it closely in to the chest wall), facilitating pushing movements such as press-ups or punching.

Nerve
Long thoracic nerve, C**5**, **6**, **7**, **8**.

Artery
Lateral thoracic artery
(from axillary artery).

Basic functional movement
Example: Reaching forwards for something barely within reach.

Note: A lesion of the long thoracic nerve will result in the medial border of the scapula falling away from the posterior chest wall; resulting in a 'winged scapula' (looking like an angel's wing). A weak muscle will also produce a winged scapula, especially when holding a weight in front of the body.

PECTORALIS MINOR

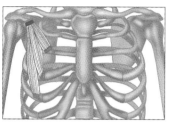

Anterior view.

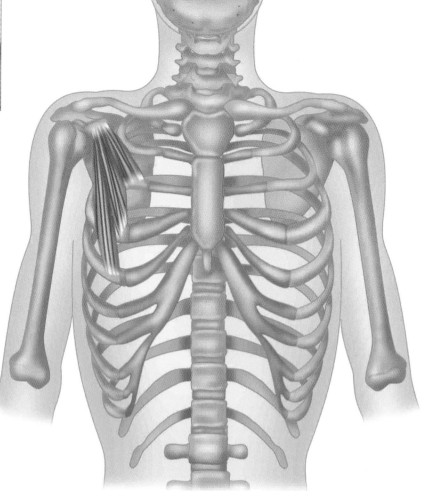

Latin, *pectoralis*, chest; *minor*, small.

Pectoralis minor is a flat triangular muscle lying posterior to, and concealed by, pectoralis major. Along with pectoralis major, it forms the anterior wall of the axilla.

Origin
Outer surfaces of third, fourth and fifth ribs and fascia of the corresponding intercostal spaces.

Insertion
Corocoid process of scapula.

Action
Draws scapula forward and downward. Raises ribs during forced inspiration (i.e. it is an accessory muscle of inspiration, if the scapula is stabilised by the rhomboids and trapezius).

Nerve
Medial pectoral nerve with fibres from a communicating branch of the lateral pectoral nerve, C(6), **7**, **8**, T1.

Artery
Pectoral branch of the thoracoacromial trunk (from axillary artery).
Can also be supplied by lateral thoracic artery.

Basic functional movement
Example: Pushing on arms of chair to stand up.

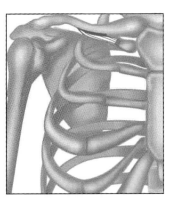

Anterior view.

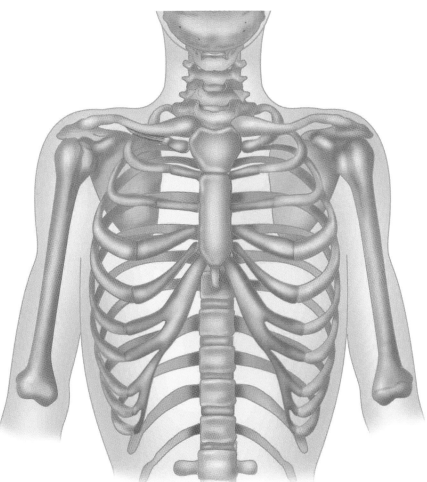

Latin, *sub*, under; *clavis*, key, clavicle.

This muscle is posterior to, and concealed by, the clavicle and pectoralis major. Paralysis of this muscle produces no apparent effect.

Origin
Junction of the first rib and the first costal cartilage.

Insertion
Floor of a groove on the lower (inferior) surface of the clavicle.

Action
Depresses clavicle and draws it towards the sternum, thereby steadying the clavicle in movements of the shoulder girdle.

Nerve
Nerve to subclavius, C**5**, **6**.

Artery
Clavicular branch of the thoracoacromial trunk (from axillary artery).

PECTORALIS MAJOR

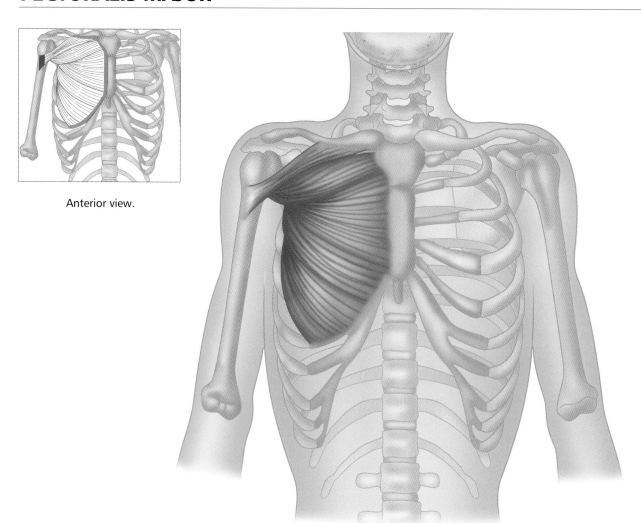

Anterior view.

Latin, *pectoralis*, chest; *major*, large.

Along with pectoralis minor, pectoralis major forms the anterior wall of the axilla.

Origin
Clavicular head: medial half or two thirds of front of clavicle.
Sternocostal portion: front of manubrium and body of sternum. Upper six costal cartilages. Rectus sheath.

Insertion
Crest below greater tubercle of humerus. Lateral lip of intertubular sulcus (bicipital groove) of humerus.

Action
Adducts and medially rotates the humerus.
Clavicular portion: flexes and medially rotates the shoulder joint, and horizontally adducts the humerus towards the opposite shoulder.
Sternocostal portion: obliquely adducts the humerus

towards the opposite hip.

The pectoralis major is one of the main climbing muscles, pulling the body up to the fixed arm.

Nerve
Nerve to upper fibres: lateral pectoral nerve C**5**, **6**, **7**.
Nerve to lower fibres: lateral and medial pectoral nerves C**6**, **7**, **8**, T**1**.

Artery
Pectoral branch of the thoracoacromial trunk and lateral thoracic artery
(from axillary artery).

Basic functional movement
Clavicular portion: Brings arm forwards and across the body, e.g. as in applying deodorant to opposite armpit.
Sternocostal portion: Pulling something down from above, e.g. such as a rope in bell ringing.

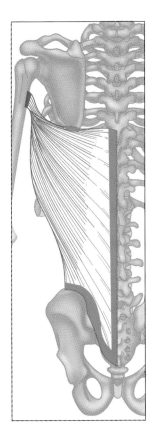

Posterior view.

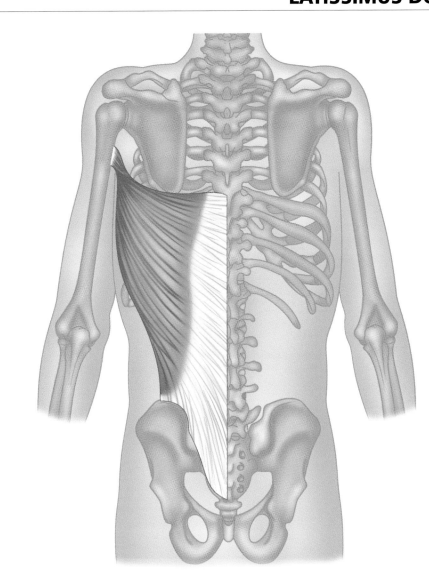

Latin, *latissimus*, widest; *dorsi*, of the back.

Along with subscapularis and teres major, the latissimus dorsi forms the posterior wall of the axilla.

Origin
Thoracolumbar fascia, which is attached to the spinous processes of lower six thoracic vertebrae and all the lumbar and sacral vertebrae (T7–S5) and to the intervening supraspinous ligaments. Posterior part of iliac crest. Lower three or four ribs. Inferior angle of the scapula.

Insertion
Floor of the intertubercular sulcus (bicipital groove) of humerus.

Action
Extends the flexed arm. Adducts and medially rotates the humerus.

It is one of the chief climbing muscles, since it pulls the shoulders downwards and backwards, and pulls the trunk up to the fixed arms (therefore, also active in crawl swimming stroke). Assists in forced inspiration, by raising the lower ribs.

Nerve
Thoracodorsal nerve, C**6**, **7**, **8**, from the posterior cord of the brachial plexus.

Artery
Thoracodorsal artery
via subscapular artery (from axillary artery).
Dorsal scapular artery
via deep branch of transverse cervical artery (from subclavian artery).

Basic functional movement
Example: Pushing on arms of chair to stand up.

MUSCLES OF THE SHOULDER JOINT

This section includes the group of muscles that run between the scapula and humerus, which act exclusively upon the shoulder joint: i.e. *deltoid, supraspinatus, infraspinatus, teres minor, subscapularis* and *teres major*. *Coracobrachialis* also acts exclusively upon the shoulder joint, but because of its position, has been included within muscles of the arm.

Deltoideus

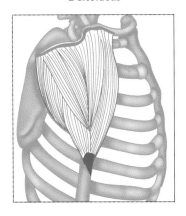

Supraspinatus

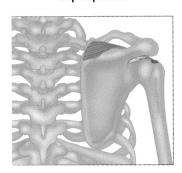

Infraspinatus

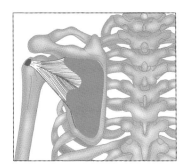

Subscapularis

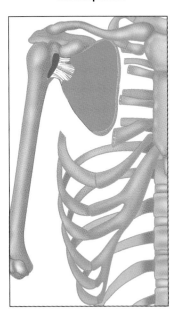

Teres minor

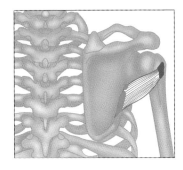

Teres major

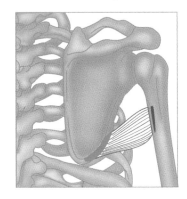

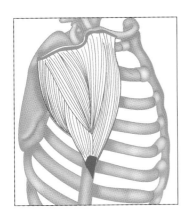

Lateral view.

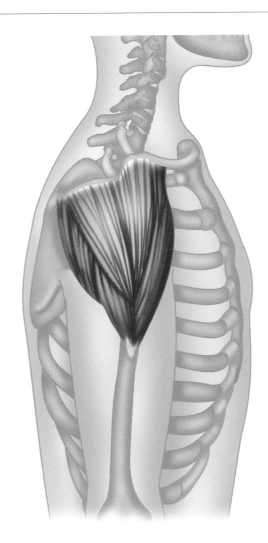

Greek, *delta*, fourth letter of Greek alphabet (shaped like a triangle).

The deltoid is composed of three parts; anterior, middle and posterior. Only the middle part is multipennate, probably because its mechanical disadvantage of abduction of the shoulder joint requires extra strength.

Origin
Anterior fibres: anterior border and superior surface of the lateral third of clavicle.
Middle fibres: lateral border of the acromion process.
Posterior fibres: lower lip of the crest of the spine of the scapula.

Insertion
Deltoid tuberosity situated half way down the lateral surface of the shaft of the humerus.

Action
Anterior fibres: flex and medially rotate the humerus.
Middle fibres: abduct the humerus at the shoulder joint (only after the movement has been initiated by supraspinatus).
Posterior fibres: extend and laterally rotate the humerus.

Nerve
Axillary nerve, C**5**, **6**, from the posterior cord of the brachial plexus.

Artery
Posterior circumflex humeral artery and deltoid branch of thoracoacromial artery (from axillary artery).

Basic functional movement
Examples: Reaching for something out to the side. Raising the arm to wave.

SUPRASPINATUS

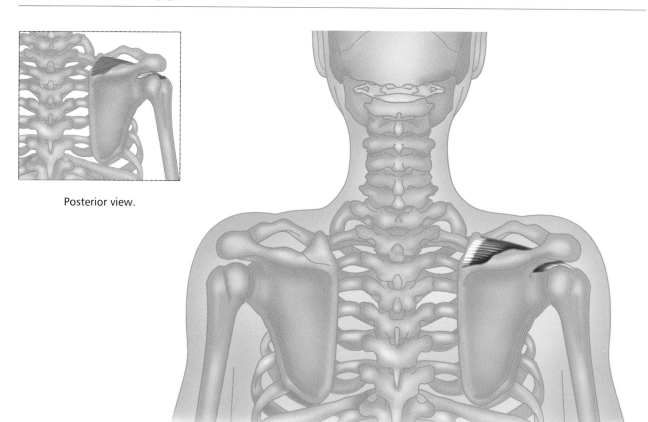

Posterior view.

Latin, *supra*, above; *spina*, spine.

A member of the **rotator cuff**, which comprise: *supraspinatus, infraspinatus, teres minor*, and *subscapularis*. The rotator cuff helps hold the head of the humerus in contact with the glenoid cavity (fossa, socket) of the scapula during movements of the shoulder, thus helping to prevent dislocation of the joint.

Origin
Supraspinous fossa of scapula.

Insertion
Upper aspect of the greater tubercle of the humerus. Capsule of shoulder joint.

Action
Initiates the process of abduction at the shoulder joint, so that the deltoid can take over at the later stages of abduction.

Nerve
Suprascapular nerve, C4, **5**, 6, from the upper trunk of the brachial plexus.

Artery
Suprascapular artery via thyrocervical trunk
(from subclavian artery).

Basic functional movement
Example: Holding shopping bag away from side of body.

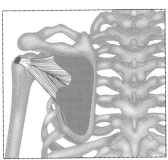

Posterior view.

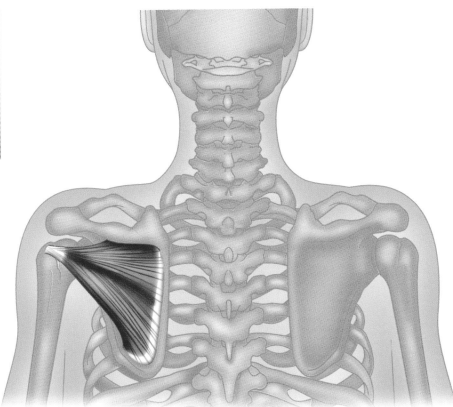

Latin, *infra,* below; *spina,* spine.

A member of the **rotator cuff,** which comprise: *supraspinatus, infraspinatus, teres minor,* and *subscapularis.* The rotator cuff helps hold the head of the humerus in contact with the glenoid cavity (fossa, socket) of the scapula during movements of the shoulder, thus helping to prevent dislocation of the joint.

Origin
Infraspinous fossa of the scapula.

Insertion
Middle facet on the greater tubercle of humerus. Capsule of shoulder joint.

Action
As a rotator cuff, helps prevent posterior dislocation of the shoulder joint. Laterally rotates humerus.

Nerve
Suprascapular nerve, C(4), **5, 6,** from the upper trunk of the brachial plexus.

Artery
Suprascapular artery
via thyrocervical trunk (from subclavian artery).
Circumflex scapula artery
via subscapular artery (from axillary artery).

Basic functional movement
Example: Brushing back of hair.

TERES MINOR

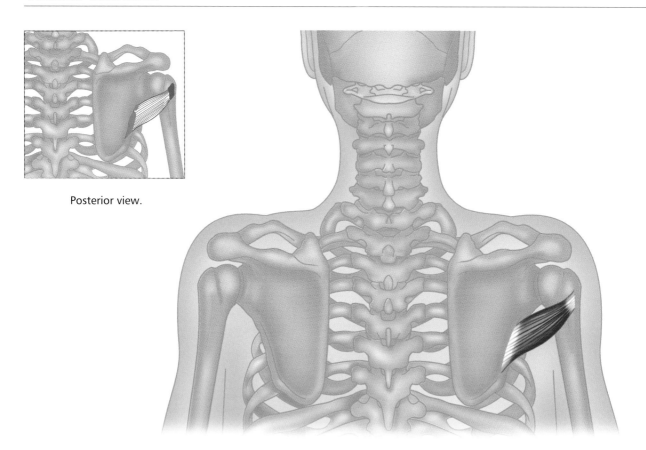

Posterior view.

Latin, *teres,* rounded; finely shaped; *minor,* small.

A member of the **rotator cuff**, which comprise: *supraspinatus, infraspinatus, teres minor,* and *subscapularis.* The rotator cuff helps hold the head of the humerus in contact with the glenoid cavity (fossa, socket) of the scapula during movements of the shoulder, thus helping to prevent dislocation of the joint.

Origin
Upper two-thirds of the lateral border of the dorsal surface of scapula.

Insertion
Lower facet on the greater tubercle of humerus. Capsule of shoulder joint.

Action
As a rotator cuff, helps prevent upward dislocation of the shoulder joint. Laterally rotates humerus. Weakly adducts humerus.

Nerve
Axillary nerve, C**5, 6**, from the posterior cord of the brachial plexus.

Artery
Circumflex scapula artery
via subscapular artery (from axillary artery).

Basic functional movement
Example: Brushing back of hair.

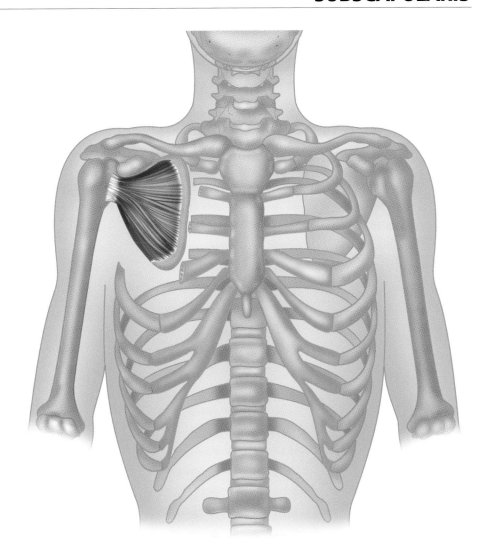

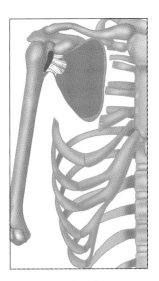

Anterior view.

Latin, *sub*, under; *scapular*, pertaining to the scapula.

A member of the **rotator cuff**, which comprise: *supraspinatus, infraspinatus, teres minor,* and *subscapularis*. The rotator cuff helps hold the head of the humerus in contact with the glenoid cavity (fossa, socket) of the scapula during movements of the shoulder, thus helping to prevent dislocation of the joint. The subscapularis constitutes the greater part of the posterior wall of the axilla.

Origin
Subscapular fossa and the groove along the lateral border of the anterior surface of scapula.

Insertion
Lesser tubercle of humerus. Capsule of shoulder joint.

Action
As a rotator cuff, stabilizes glenohumeral joint; mainly preventing the head of the humerus being pulled upwards by the deltoid, biceps and long head of triceps. Medially rotates humerus.

Nerve
Upper and lower subscapular nerves, C**5**, **6**, 7, from the posterior cord of the brachial plexus.

Artery
Subscapular artery
(from axillary artery).

Basic functional movement
Example: Reaching into your back pocket.

TERES MAJOR

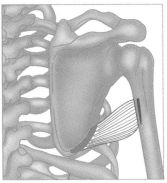

Posterior view.

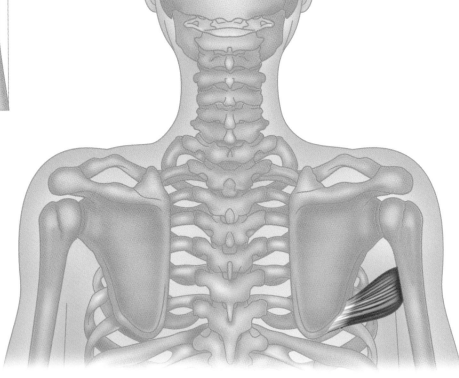

Latin, *teres,* rounded, finely shaped; *major,* large.

The teres major, along with the tendon of latissimus dorsi, which passes around it, and the subscapularis, forms the posterior fold of the axilla.

Origin
Oval area on the lower third of the posterior surface of the lateral border of the scapula.

Insertion
Medial lip of the intertubercular sulcus (bicipital groove) of humerus.

Action
Adducts humerus. Medially rotates humerus. Extends humerus from the flexed position.

Nerve
Lower subscapular nerve, C5, **6**, 7, from the posterior cord of the brachial plexus.

Artery
Circumflex scapula artery
via subscapular artery (from axillary artery).

Basic functional movement
Example: Reaching into your back pocket.

Muscles of the arm comprise those that originate from the scapula and / or the humerus, and insert into the radius and / or ulna; so that they act upon the elbow joint. These are: *biceps brachii, brachialis, triceps brachii,* and *anconeus. Coracobrachialis* which, although acting upon the shoulder joint, is also included here because of its proximity to the other muscles of this group.

Biceps brachii

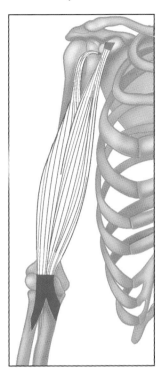

Brachialis

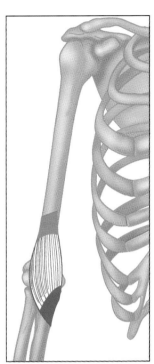

Anconeus

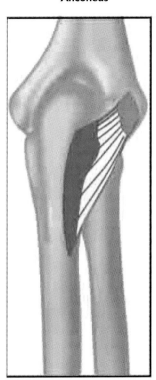

Coracobrachialis

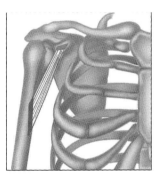

Triceps brachii

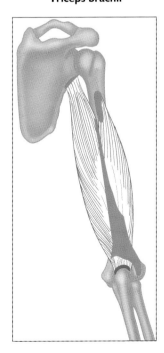

BICEPS BRACHII

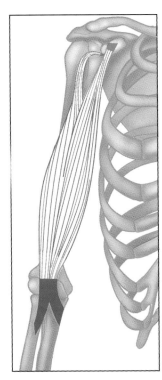

Anterior view, right arm.

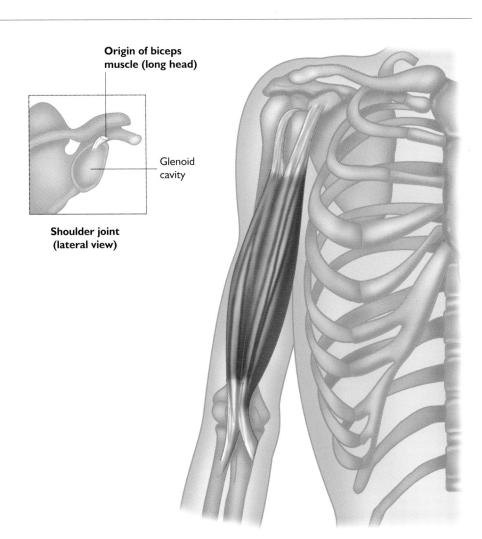

Origin of biceps muscle (long head)

Glenoid cavity

Shoulder joint (lateral view)

Latin, *biceps*, two headed muscle; *brachii*, of the arm.

Biceps brachii operates over three joints. It has two tendinous heads at its origin and two tendinous insertions. Occasionally it has a third head, originating at the insertion of coracobrachialis. The short head forms part of the lateral wall of the axilla, along with coracobrachialis and the humerus.

Origin
Short head: tip of corocoid process of scapula.
Long head: supraglenoid tubercle of scapula.

Insertion
Posterior part of radial tuberosity.
Bicipital aponeurosis, which leads into the deep fascia on medial aspect of forearm.

Action
Flexes elbow joint. Supinates forearm. (It has been described as the muscle that puts in the corkscrew and pulls out the cork). Weakly flexes arm at the shoulder joint.

Nerve
Musculocutaneous nerve, C**5**, **6**.

Artery
Brachial artery
(a continuation of the axillary artery).

Basic functional movement
Examples: Picking up an object. Bringing food to mouth.

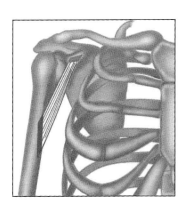

Anterior view, right arm.

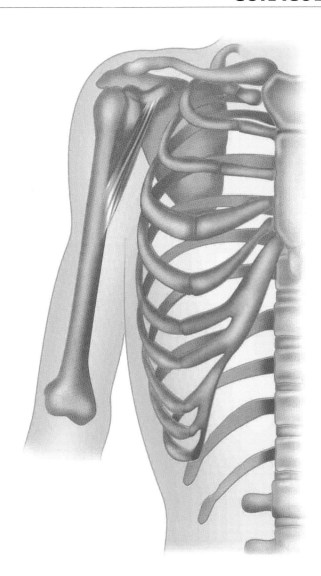

Greek, *coracoid*, raven's beak; **Latin**, *brachial*; relating to the arm.

Along with the short head of biceps brachii and the humerus, the coracobrachialis forms the lateral wall of the axilla.

Origin
Tip of the corocoid process of scapula.

Insertion
Medial aspect of humerus at mid-shaft.

Action
Weakly adducts shoulder joint. Possibly assists in flexion of the shoulder joint, (but this has not been proven). Helps stabilize humerus.

Nerve
Musculocutaneous nerve, C**6, 7**.

Artery
Brachial artery
(a continuation of the axillary artery).

Basic functional movement
Example: Mopping the floor.

BRACHIALIS

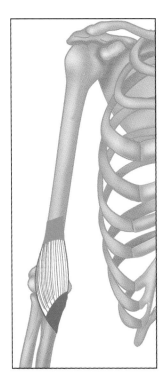

Anterior view, right arm.

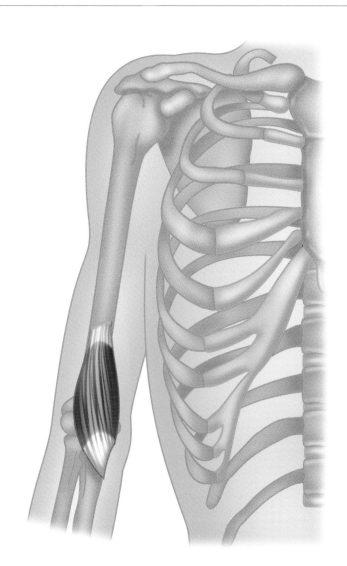

Latin, *brachial*, relating to the arm.

Brachialis lies posterior to biceps brachii and is the main flexor of the elbow joint. Some fibres may be partly fused with the brachioradialis.

Origin
Lower (distal) two thirds of anterior aspect of humerus.

Insertion
Coronoid process of ulna and tuberosity of ulna, (i.e. area on front of upper part of shaft of ulna).

Action
Flexes elbow joint.

Nerve
Musculocutaneous nerve, C**5**, **6**.

Artery
Brachial artery
(a continuation of the axillary artery).
Radial recurrent artery
(from radial artery).

Basic functional movement
Example: Bringing food to the mouth.

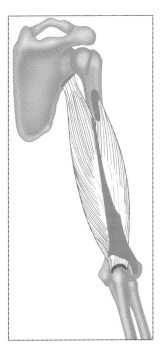

Posterior view, right arm.

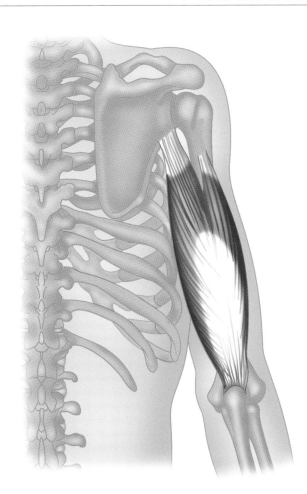

Latin, *triceps*, three headed muscle; *brachii*, of the arm.

The triceps originates from three heads and is the only muscle on the back of the arm.

Origin
Long head: infraglenoid tubercle of the scapula.
Lateral head: upper half of posterior surface of shaft of humerus (above and lateral to the radial groove).
Medial head: lower half of posterior surface of shaft of humerus (below and medial to the radial groove).

Insertion
Posterior part of the olecranon process of the ulna.

Action
Extends elbow joint. Long head can adduct the humerus and extend it from the flexed position. Stabilizes shoulder joint.

Nerve
Radial nerve, C6, **7**, **8**, T1.

Artery
Deep brachial artery
(via brachial artery continuing from axillary artery).

Basic functional movement
Examples: Throwing objects. Pushing a door shut.

ANCONEUS

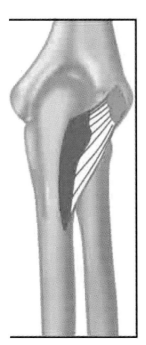

Posterior view, right arm.

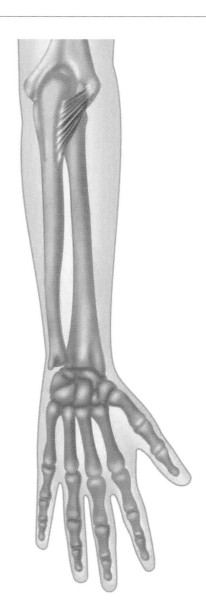

Greek, elbow.

Origin
Posterior part of lateral epicondyle of humerus.

Insertion
Lateral surface of the olecranon process and upper portion of posterior surface of ulna.

Action
Assists triceps to extend forearm at elbow joint. May stabilize the ulna during pronation and supination.

Nerve
Radial nerve, C7, **8.**

Artery
Middle collateral branch of deep brachial artery
(from ulnar artery).
Recurrent interosseous artery
via common interosseous artery (from ulnar artery).

Basic functional movement
Example: Pushing objects at arms length.

Muscles of the Forearm and Hand

12

MUSCLES OF THE ANTERIOR FOREARM

The anterior forearm contains three functional muscle groups: the pronators of the forearm, the wrist flexors, and the long flexors of the fingers and thumb. They are arranged in three layers: the superficial layer comprises four muscles, all emanating from a common tendon known as the common flexor origin. These are: *pronator teres, flexor carpi radialis, palmaris longus,* and *flexor carpi ulnaris.* The middle layer contains only the *flexor digitorum superficialis.* The deepest layer consists of: *flexor digitorum profundus, flexor pollicis longus* and *pronator quadratus.*

Pronator teres

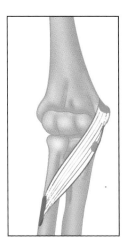

Palmaris longus

Flexor digitorum superficialis

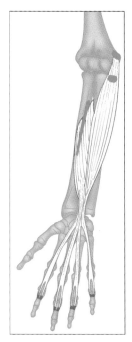

Flexor pollicis longus

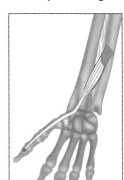

Flexor carpi radialis

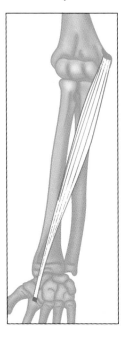

Flexor carpi ulnaris

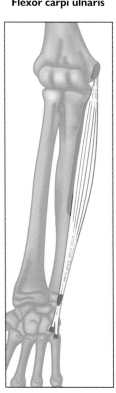

Flexor digitorum profundus

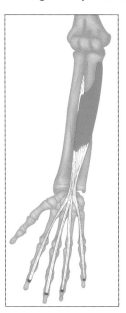

Pronator quadratus

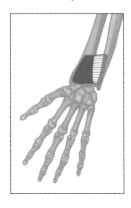

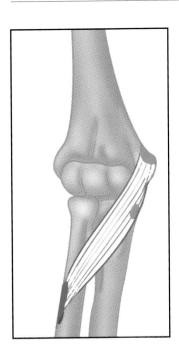

Anterior view, right arm.

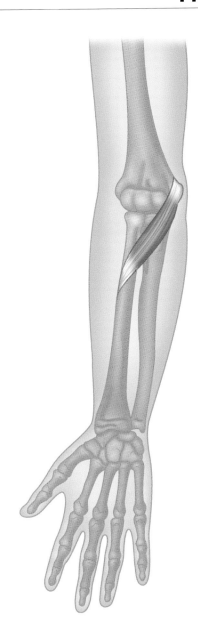

Latin, *pronate*, bent forward; *teres*, rounded, finely shaped.

Part of the superficial layer, which includes: pronator teres, flexor carpi radialis, palmaris longus, and flexor carpi ulnaris.

Origin
Humeral head: lower third of medial supracondylar ridge and the common flexor origin on the anterior aspect of the medial epicondyle of humerus.
Ulnar head: medial border of the coronoid process of the ulna.

Insertion
Mid-lateral surface of radius (pronator tuberosity).

Action
Pronates forearm. Assists flexion of elbow joint.

Nerve
Median nerve, C6, 7.

Artery
Ulnar artery
(from brachial artery).
Anterior ulnar recurrent branch of ulnar artery

Basic functional movement
Examples: Pouring liquid from a container. Turning a doorknob.

FLEXOR CARPI RADIALIS

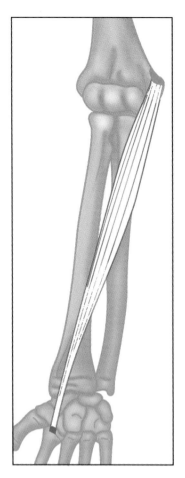

Anterior view, right arm.

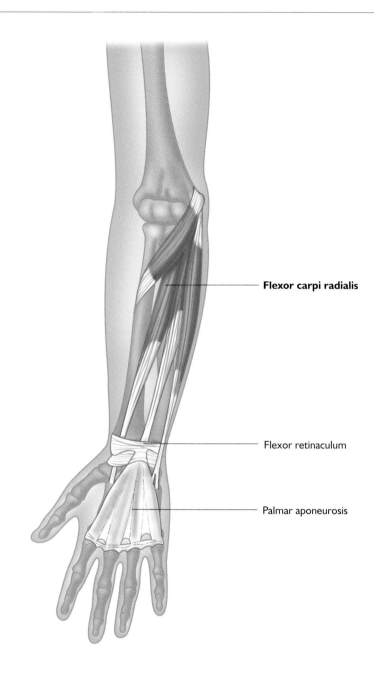

Flexor carpi radialis

Flexor retinaculum

Palmar aponeurosis

Latin, *flex*, to bend; *carpi*, of the wrist; *radius*, staff, spoke of wheel.

Part of the superficial layer, which includes: pronator teres, flexor carpi radialis, palmaris longus, and flexor carpi ulnaris.

Origin
Common flexor origin on the anterior aspect of the medial epicondyle of humerus.

Insertion
Front of the bases of the second and third metacarpal bones.

Action
Flexes and abducts the carpus (wrist joint). Helps to flex the elbow and pronate the forearm.

Nerve
Median nerve, C**6**, **7**, 8.

Artery
Ulnar and radial arteries
(from brachial artery).

Basic functional movement
Examples: Pulling rope in towards you. Wielding an axe or hammer.

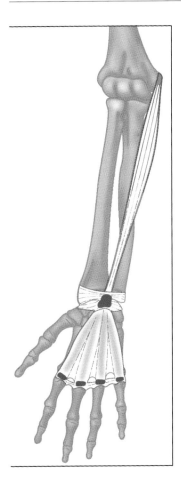

Anterior view, right arm.

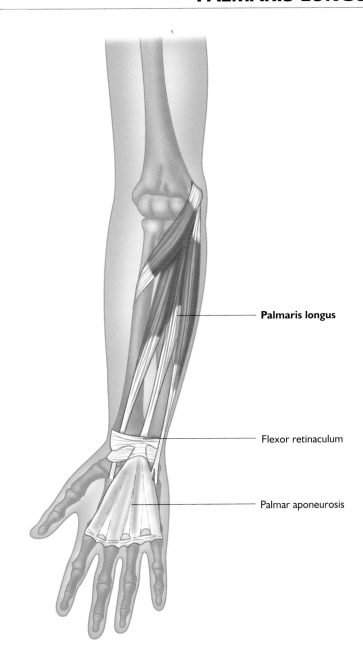

Palmaris longus

Flexor retinaculum

Palmar aponeurosis

Latin, *palmaris,* palma, palm; *longus,* long.

Part of the superficial layer, which includes: pronator teres, flexor carpi radialis, palmaris longus, and flexor carpi ulnaris. The palmaris longus muscle is frequently absent.

Origin
Common flexor origin on the anterior aspect of the medial epicondyle of humerus.

Insertion
Superficial (front) surface of flexor retinaculum and apex of the palmar aponeurosis.

Action
Flexes the wrist. Tenses the palmar fascia.

Nerve
Median nerve C(6), **7**, **8**, T1.

Artery
Ulnar artery
(from brachial artery).

Basic functional movement
Examples: Grasping a small ball. Cupping the palm to drink from the hand.

FLEXOR CARPI ULNARIS

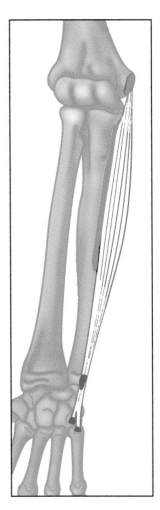

Anterior view, right arm.

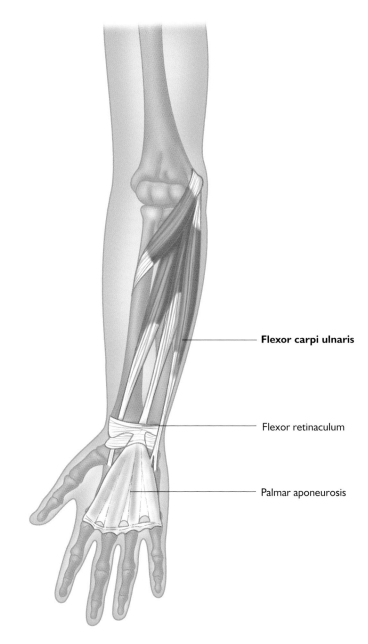

Flexor carpi ulnaris

Flexor retinaculum

Palmar aponeurosis

Latin, *flex*, to bend; *carpi*, of the wrist; *ulnaris*, of the elbow / arm.

Part of the superficial layer, which includes: pronator teres, flexor carpi radialis, palmaris longus, and flexor carpi ulnaris.

Origin
Humeral head: common flexor origin on the medial epicondyle of humerus.
Ulnar head: medial border of olecranon. Posterior border of upper two-thirds of ulna.

Insertion
Pisiform bone. Hook of hamate. Base of fifth metacarpal.

Action
Flexes and adducts the wrist. May weakly assist in flexion of elbow.

Nerve
Ulnar nerve, C7, **8**, T1.

Artery
Ulnar artery
(from brachial artery).

Basic functional movement
Example: Pulling an object towards you.

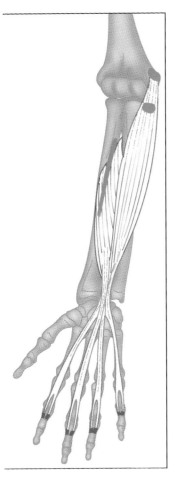

Anterior view, right arm.

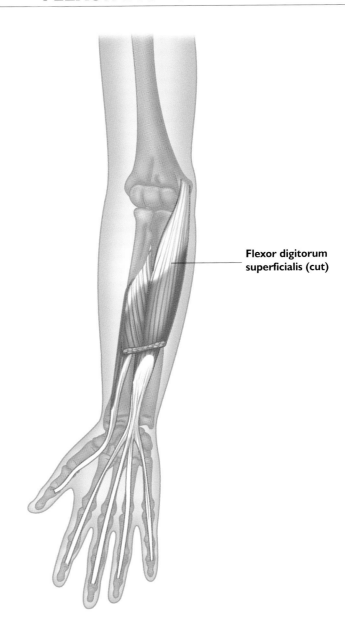

Flexor digitorum superficialis (cut)

Latin, *flex*, to bend; *digit*, finger; *superficialis*, on the surface.

This muscle alone constitutes the middle layer of anterior forearm muscles.

Origin
Humeroulnar head: long linear origin from common flexor tendon on medial epicondyle of humerus. Medial border of coronoid process of ulna.
Radial head: upper two-thirds of anterior border of radius.

Insertion
Four tendons each divide into two slips, each of which insert into the sides of the middle phalanges of the four fingers.

Action
Flexes the middle phalanges of each finger. Can help flex the wrist.

Nerve
Median nerve, C**7**, **8**, T1.

Artery
Ulnar artery
(from brachial artery).

Basic functional movement
Examples: 'Hook grip', 'power grip', as in turning a tap, typing, playing the piano and some stringed instruments.

FLEXOR DIGITORUM PROFUNDUS

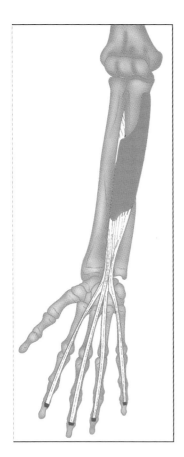

Anterior view, right arm.

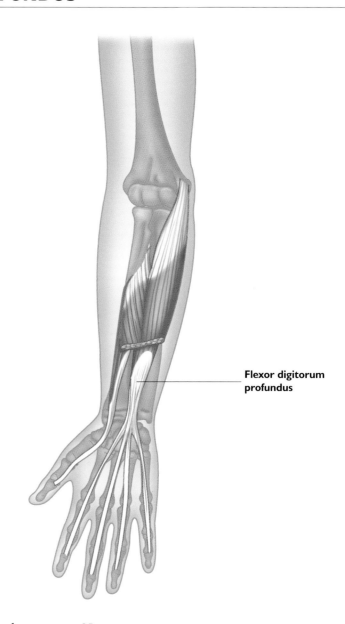

Flexor digitorum profundus

Latin, *flex,* to bend; *digit,* finger; *profundus,* deep.

Part of the deep layer (third layer), which includes: flexor digitorum profundus, flexor pollicis longus, and pronator quadratus. In the palm, the tendons of flexor digitorum profundus give origin to the lumbricales.

Origin
Upper two-thirds of the medial and anterior surfaces of the ulna, reaching up onto the medial side of the olecranon process. Interosseous membrane.

Insertion
Anterior surface of base of distal phalanges.

Action
Flexes distal phalanges (the only muscle able to do so). Helps flex all joints across which it passes.

Nerve
Medial half of muscle, destined for the little and ring fingers: ulnar nerve, C7, **8**, T**1**.
Lateral half of muscle, destined for the index and middle fingers: anterior interosseous branch of median nerve, C7, **8**, T1. Sometimes the ulnar nerve supplies the whole muscle.

Artery
Ulnar artery
(from brachial artery).
Anterior interosseous artery
(from ulnar artery).

Basic functional movement
Example: 'Hook' grip, as in carrying a briefcase.

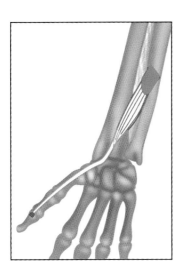

Anterior view, right arm.

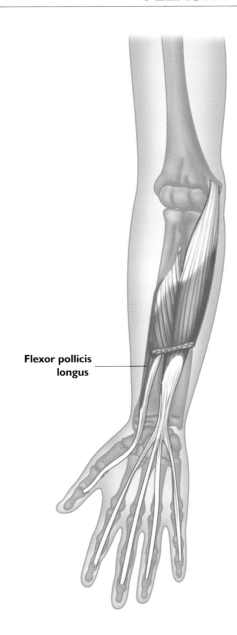

Flexor pollicis longus

Latin, *flex,* to bend; *pollicis,* of the thumb; *longus,* long.

Part of the deep layer (third layer), which includes: flexor digitorum profundus, flexor pollicis longus, and pronator quadratus. Its tendon, along with those of the other long digital flexor tendons, passes through the carpal tunnel.

Origin
Middle part of anterior surface of shaft of radius. Interosseous membrane. Medial border of coronoid process of ulna and/or medial epicondyle of humerus.

Insertion
Palmar surface of base of distal phalanx of thumb.

Action
Flexes the interphalangeal joint of the thumb (the only muscle able to do so). Assists in flexion of the metacarpophalangeal and carpometacarpal joints. Can assist in flexion of the wrist.

Nerve
Anterior interosseous branch of median nerve, C(6), 7, **8**, T1.

Artery
Anterior interosseous artery
(from ulnar artery).

Basic functional movement
Examples: Picking up small objects between thumb and fingers. Maintaining firm grip on a hammer.

PRONATOR QUADRATUS

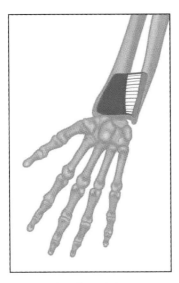

Anterior view, right arm.

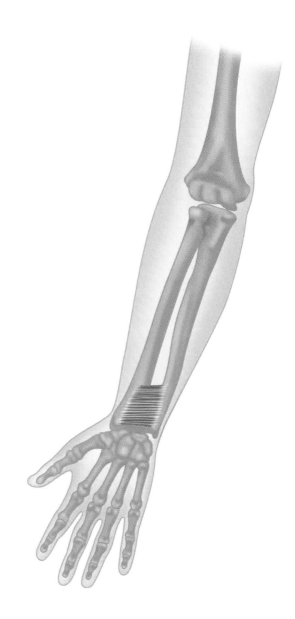

Latin, *pronate*, bent forward; *quadratus*, squared.

Part of the deep layer (third layer), which includes: flexor digitorum profundus, flexor pollicis longus, and pronator quadratus.

Origin
Distal quarter of anterior surface of shaft of ulna.

Insertion
Lateral side of distal quarter of anterior surface of shaft of radius.

Action
Pronates forearm and hand. Helps hold radius and ulna together, reducing stress on the inferior radio-ulnar joint.

Nerve
Anterior interosseous branch of median nerve, C7, **8**, T**1**.

Artery
Anterior interosseous artery
(from ulnar artery).

Basic functional movement
Example: Turning hand downwards as in pouring a substance out of the hand.

MUSCLES OF THE POSTERIOR FOREARM

On the back of the forearm there are two muscle groups. The superficial group contains, from the radial to ulnar side: brachioradialis, extensor carpi radialis longus, extensor carpi radialis brevis, extensor digitorum, extensor digiti minimi and extensor carpi ulnaris. The deep group contains: supinator, abductor pollicis longus, extensor pollicis brevis, extensor pollicis longus, and extensor indicis.

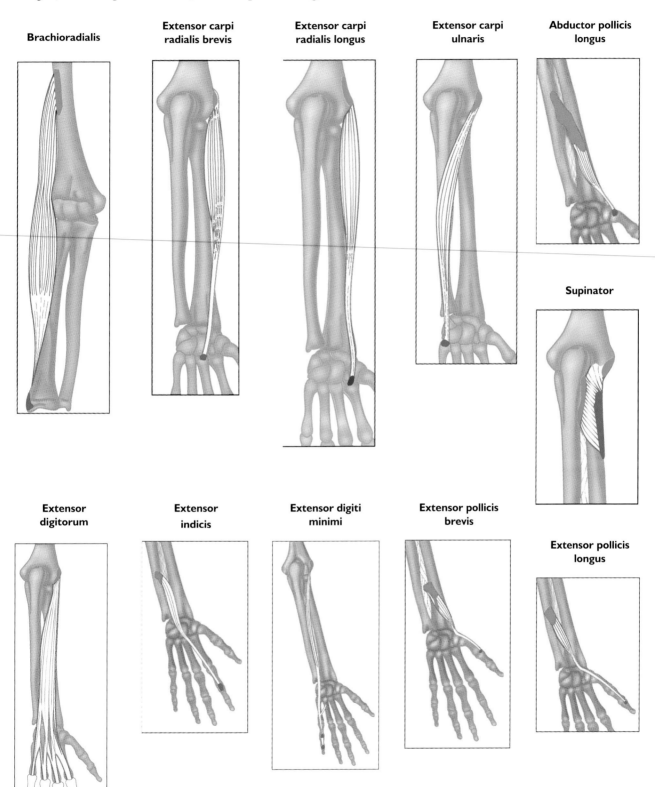

Brachioradialis

Extensor carpi radialis brevis

Extensor carpi radialis longus

Extensor carpi ulnaris

Abductor pollicis longus

Supinator

Extensor digitorum

Extensor indicis

Extensor digiti minimi

Extensor pollicis brevis

Extensor pollicis longus

BRACHIORADIALIS

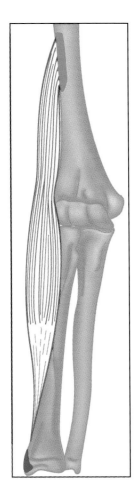

Anterior view,
right arm.

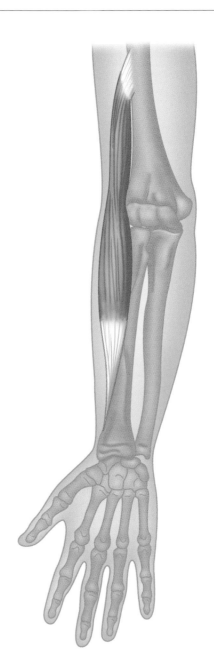

Latin, *brachial*, relating to the arm; *radius*, staff, spoke of wheel.

Part of the superficial group. The brachioradialis forms the lateral border of the cubital fossa. The muscle belly is prominent when working against resistance.

Origin
Upper two-thirds of the anterior aspect of lateral supracondylar ridge of humerus.

Insertion
Lower lateral end of radius, just above the styloid process.

Action
Flexes elbow joint. Assists in pronating and supinating forearm when these movements are resisted.

Nerve
Radial nerve, C**5**, **6**.

Artery
Radial recurrent branch of radial artery (from brachial artery).

Basic functional movement
Example: Turning a corkscrew.

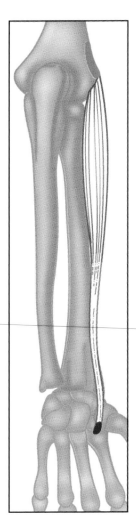

Posterior view,
right arm.

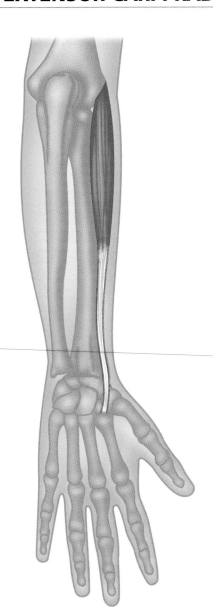

Latin, *extensor*, to extend; *carpi*, of the wrist; *radius*, staff, spoke of wheel; *longus*, long.

Part of the superficial group. The fibres of this muscle are often blended with those of brachioradialis.

Origin
Lower (distal) third of lateral supracondylar ridge of humerus.

Insertion
Dorsal surface of base of second metacarpal bone, on its radial side.

Action
Extends and abducts the wrist. Assists in flexion of the elbow.

Nerve
Radial nerve, C5, **6**, **7**, 8.

Artery
Radial artery
(from brachial artery).

Basic functional movement
Examples: Kneading dough. Typing.

EXTENSOR CARPI RADIALIS BREVIS

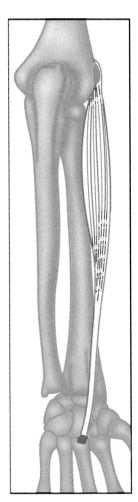

Posterior view,
right arm.

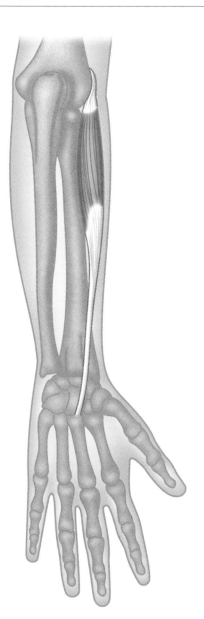

Latin, *extensor*, to extend; *carpi*, of the wrist; *radius*, staff, spoke of wheel; *brevis*, short.

Part of the superficial group. This muscle is often fused with extensor carpi radialis longus at its origin.

Origin
Common extensor tendon from lateral epicondyle of humerus.

Insertion
Dorsal surface of third metacarpal.

Action
Extends wrist. Assists abduction of wrist.

Nerve
Radial nerve, C5, **6**, **7**, 8.

Artery
Radial artery
(from brachial artery).

Basic functional movement
Examples: Kneading dough. Typing.

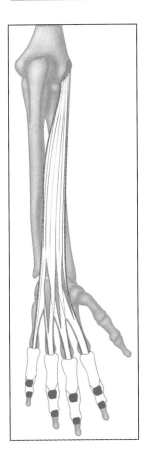

Posterior view,
right arm.

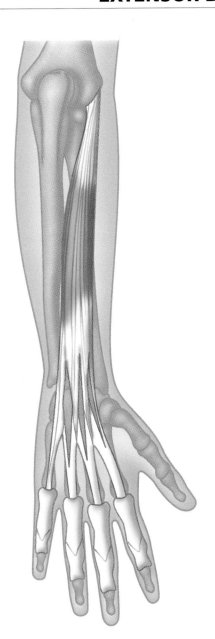

Latin, *extensor*, to extend; *digit*, finger.

Part of the superficial group. Each tendon of extensor digitorum, over each metacarpophalangeal joint, forms a triangular membranous sheet called the **extensor hood** or **extensor expansion**, into which inserts the lumbricales and interossei of the hand. Extensor digiti minimi and extensor indicis also insert into the extensor expansion.

Origin
Common extensor tendon from lateral epicondyle of humerus.

Insertion
Dorsal surfaces of all the phalanges of the four fingers.

Action
Extends the fingers (metacarpophalangeal and interphalangeal joints). Assists abduction (divergence) of fingers away from the middle finger.

Nerve
Deep radial (posterior interosseous) nerve, C**6**, **7**, **8**.

Artery
Recurrent interosseous artery and the posterior interosseous artery
via common interosseous artery (from ulnar artery).

Basic functional movement
Example: Letting go of objects held in the hand.

EXTENSOR DIGITI MINIMI

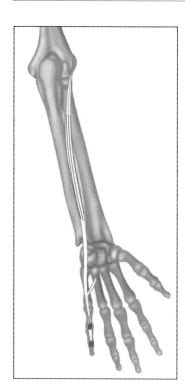

Posterior view, right arm.

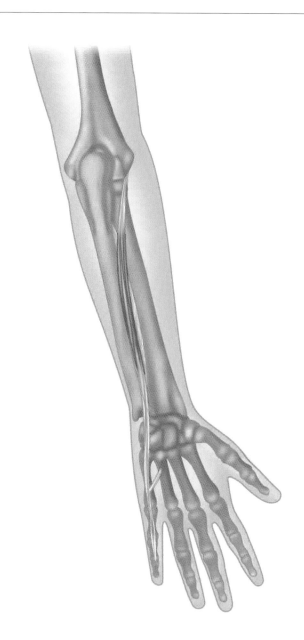

Latin, *extensor*, to extend; *digit*, finger; *minimi*, smallest.

Part of the superficial group, along with brachioradialis, extensor carpi radialis longus, extensor carpi radialis brevis, extensor digitorum, and extensor carpi ulnaris.

Origin
Common extensor tendon from lateral epicondyle of humerus.

Insertion
Extensor expansion of little finger with extensor digitorum tendon.

Action
Extends little finger.

Nerve
Deep radial (posterior interosseous) nerve, C6, **7**, **8**.

Artery
Recurrent interosseous artery
via common interosseous artery (from ulnar artery).

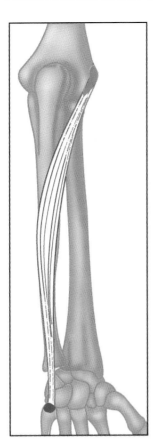

Posterior view, right arm.

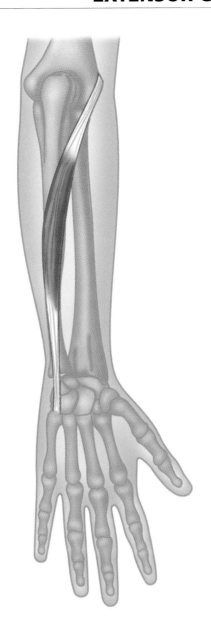

Latin, *extensor*, to extend; *carpi*, of the wrist; *ulnaris*, of the elbow.

Part of the superficial group, along with brachioradialis, extensor carpi radialis longus, extensor carpi radialis brevis, extensor digitorum, and extensor digiti minimi.

Origin
Common extensor tendon from lateral epicondyle of humerus. Aponeurosis from mid-posterior border of ulna.

Insertion
Medial side of base of fifth metacarpal.

Action
Extends and adducts the wrist.

Nerve
Deep radial (posterior interosseous) nerve, C6, **7**, **8**.

Artery
Ulnar artery
(from brachial artery).

Basic functional movement
Example: Cleaning windows.

SUPINATOR

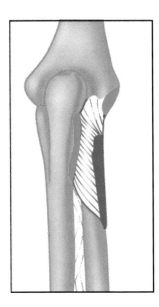

Posterior view, right arm.

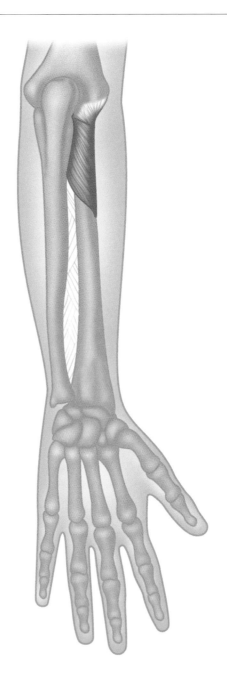

Latin, *supinus*, lying on the back.

Part of the deep group. Supinator is almost entirely concealed by the superficial muscles.

Origin
Lateral epicondyle of humerus. Radial collateral (lateral) ligament of elbow joint. Annular ligament of superior radio-ulnar joint. Supinator crest of ulna.

Insertion
Dorsal and lateral surfaces of upper third of radius.

Action
Supinates forearm (for which it is probably the main prime mover; with biceps brachii being an auxiliary).

Nerve
Deep radial nerve, C5, **6**, (7).

Artery
Recurrent interosseous artery
via common interosseous artery (from ulnar artery). Occasionally also supplied by recurrent radial artery.

Basic functional movement
Example: Turning a door handle, or screwdriver.

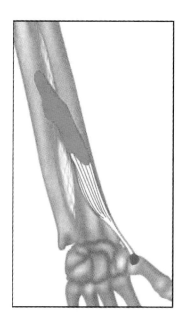

Posterior view, right arm.

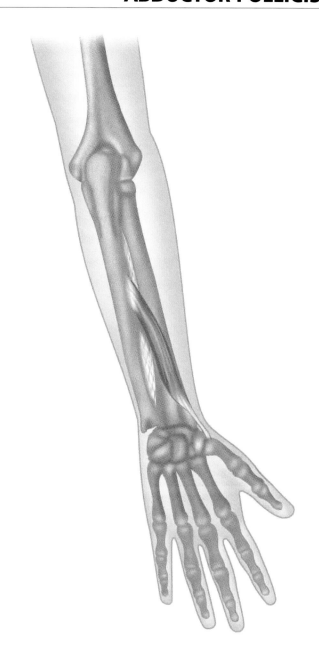

Latin, *abduct*, away from; *pollicis*, of the thumb; *longus*, long.

Part of the deep group. However, it becomes superficial in the distal part of the forearm.

Origin
Posterior surface of shaft of ulna, distal to the origin of supinator. Interosseous membrane. Posterior surface of middle third of shaft of radius.

Insertion
Radial (lateral) side of base of first metacarpal.

Action
Pulls metacarpal bone of the thumb into a position midway between extension and abduction (the tendon stands out during this movement). Abducts and assists in flexion of the wrist.

Nerve
Deep radial (posterior interosseous) nerve, C6, **7**, **8**.

Artery
Posterior interosseous artery
via common interosseous artery (from ulnar artery).

Basic functional movement
Example: Releasing grip on a flat object.

EXTENSOR POLLICIS BREVIS

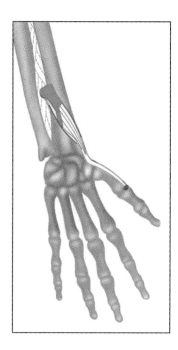

Posterior view, right arm.

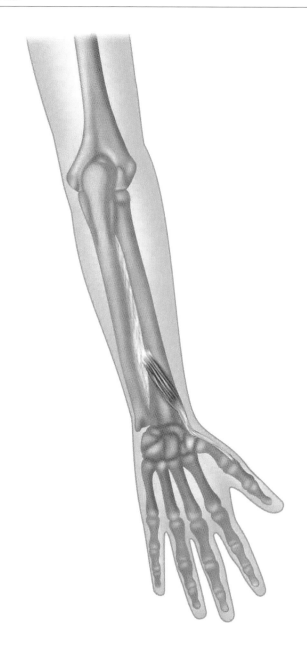

Latin, *extensor,* to extend; *pollicis,* of the thumb; *brevis,* short.

Part of the deep group. Lies distal to abductor pollicis longus, to which it closely adheres.

Origin
Posterior surface of radius, distal to origin of abductor pollicis longus. Adjacent part of interosseous membrane.

Insertion
Base of dorsal surface of proximal phalanx of thumb.

Action
Extends the thumb. Abducts the wrist.

Nerve
Deep radial (posterior interosseous) nerve, C6, **7, 8.**

Artery
Posterior interosseous artery
via common interosseous artery (from ulnar artery).

Basic functional movement
Example: Releasing grip on a flat object.

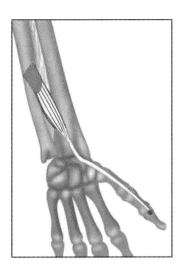

Posterior view, right arm.

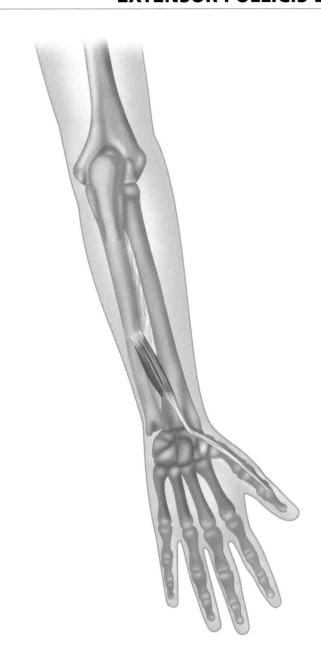

Latin, *extensor*, to extend; *pollicis*, of the thumb; *longus*, long.

Part of the deep group. The tendon of extensor pollicis longus forms the posterior boundary of the triangular hollow known as the **anatomical snuff box**, on the back of the hand, distal to the distal end of the radius.

Origin
Middle third of posterior surface of ulna. Interosseous membrane.

Insertion
Dorsal surface of base of distal phalanx of thumb.

Action
Extends thumb. Assists in extension and abduction of the wrist.

Nerve
Deep radial (posterior interosseous) nerve, C6, **7**, **8**.

Artery
Posterior interosseous artery
via common interosseous artery (from ulnar artery).

Basic functional movement
Example: Giving the 'thumbs up' gesture.

EXTENSOR INDICIS

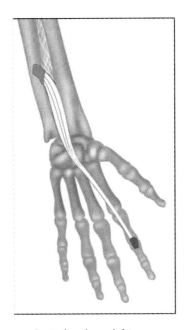

Posterior view, right arm.

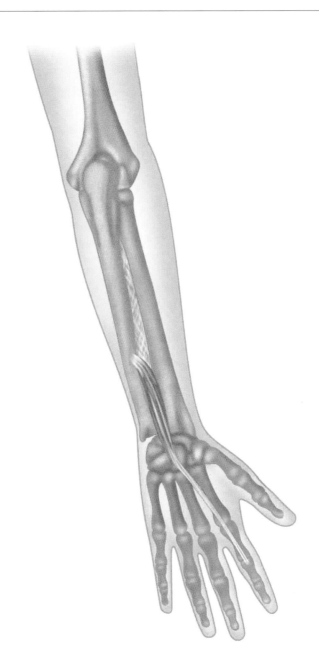

Latin, *extensor*, to extend; *indicis*, of the index.

Part of the deep group, which comprises: supinator, abductor pollicis longus, extensor pollicis brevis, extensor pollicis longus, and extensor indicis.

Origin
Posterior surface of ulna. Adjacent part of interosseous membrane.

Insertion
Extensor expansion (hood) on the dorsum of the proximal phalanx of the index finger.

Action
Extends index finger.

Nerve
Deep radial (posterior interosseous) nerve, C6, **7**, **8**.

Artery
Posterior interosseous artery
via common interosseous artery (from ulnar artery).

Basic functional movement
Example: Pointing at something.

The muscle groupings in the hand are: 1) The 'intrinsic' muscles, consisting of the *interossei*, located within the intermetacarpal spaces to act on the four fingers and thumb, and the *lumbricales*, which arise from the tendons of flexor digitorum profundus in the palm and act on the four fingers. 2) The muscles of the *hypothenar eminence*. 3) The muscles of the *thenar eminence*. 4) *Adductor pollicis*.

Lumbricales

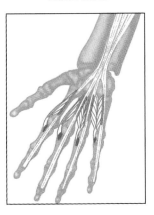

Abductor digiti minimi

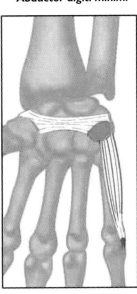

Flexor digiti minimi brevis

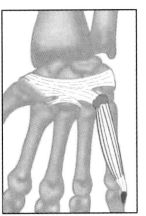

Opponens pollicis

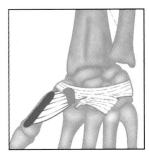

Palmar interossei

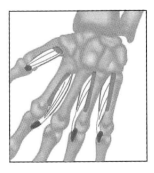

Flexor pollicis brevis

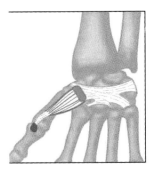

Palmaris brevis

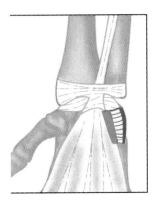

Opponens digiti minimi

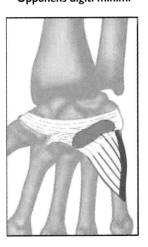

Dorsal interossei

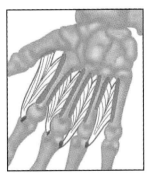

Abductor pollicis brevis

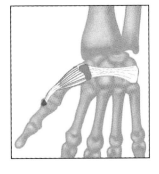

Adductor pollicis

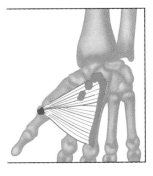

LUMBRICALES

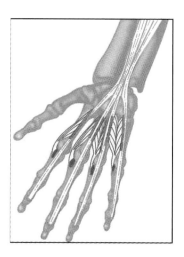

Palmar view, right hand.

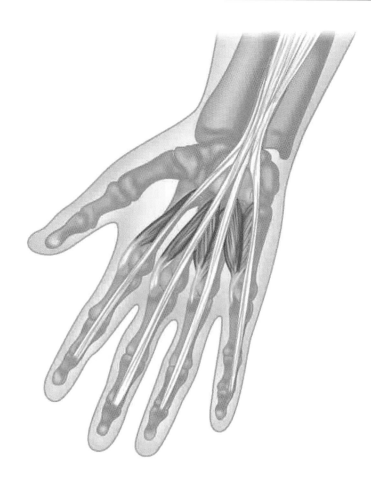

Latin, *lumbrical*, earthworm.

Four small cylindrical muscles, one for each finger, named after the earthworm, because of their shape.

Origin
Tendons of flexor digitorum profundus in the palm.

Insertion
Lateral (radial) side of corresponding tendon of extensor digitorum, on the dorsum of the respective digits.

Action
Extend the interphalangeal joints and simultaneously flex the metacarpophalangeal joints of the fingers.

Nerve
This varies. The usual configuration is:
Lateral lumbricales (first and second): median nerve, C(6), 7, **8**, T**1**.
Medial lumbricales (third and fourth): ulnar nerve, C(7), **8**, T**1**.
However, the number of lumbricales supplied by the ulnar nerve may be increased to four or decreased to one.

Artery
Palmar metacarpal arteries of deep palmar arch.

Basic functional movement
Example: Cupping your hand.

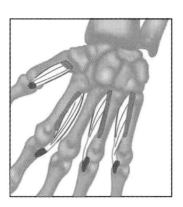

Palmar view, right hand.

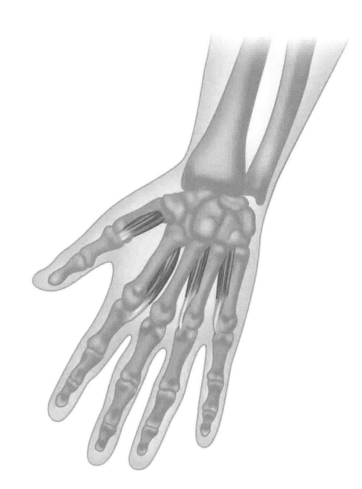

Latin, *palmaris*, palma, palm; *interosseus*, between bones.

The four palmar interossei are located in the spaces between the metacarpals. Each muscle arises from the metacarpal of the digit upon which it acts.

Origin
First: medial (ulnar) side of base of first metacarpal.
Second: medial (ulnar) side of shaft of second metacarpal.
Third: lateral (radial) side of shaft of fourth metacarpal.
Fourth: lateral (radial) side of shaft of fifth metacarpal.

Insertion
Primarily into the extensor expansion of the respective digit, with possible attachment to base of proximal phalanx as follows:
First: medial (ulnar) side of proximal phalanx of thumb.
Second: medial (ulnar) side of proximal phalanx of index finger.
Third: lateral (radial) side of proximal phalanx of ring finger.
Fourth: lateral (radial) side of proximal phalanx of little finger.

Action
Adduct (converge) fingers and thumb towards the middle (third) finger. Assist in flexion of fingers at metacarpophalangeal joints.

Nerve
Ulnar nerve C**8**, T**1**.

Artery
Palmar metacarpal arteries of deep palmar arch.

Basic functional movement
Example: Cupping hand as if to retain water in the palm (i.e. drinking from the hand).

NOTE: The palmar interosseous of the thumb is usually absent.

DORSAL INTEROSSEI

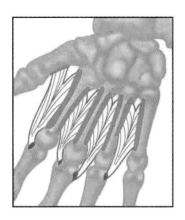

Palmar view, right hand.

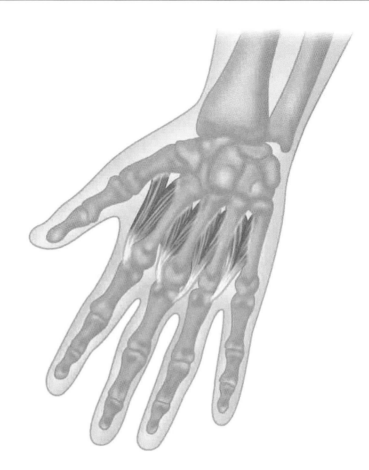

Latin, *dorsal*, back; *interosseus*, between bones.

The four dorsal interossei are about twice the size of the palmar interossei.

Origin
By two heads, each from adjacent sides of metacarpals. Therefore, each dorsal interossei occupies an interspace between adjacent metacarpals.

Insertion
Into the extensor expansion and to base of proximal phalanx as follows:
First: lateral (radial) side of index finger, mainly to base of proximal phalanx.
Second: lateral (radial) side of middle finger.
Third: medial (ulnar) side of middle finger, mainly into extensor expansion.
Fourth: medial (ulnar) side of ring finger.

Action
Abduct fingers away from middle finger. Assist in flexion of fingers at metacarpophalangeal joints.

Nerve
Ulnar nerve C**8**, T**1**.

Artery
Dorsal metacarpal arteries, and palmar metacarpal arteries of deep palmar arch.

Basic functional movement
Example: Spreading fingers, as if to indicate numbers from two to four.

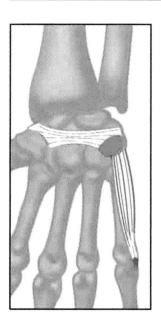

Palmar view, right hand.

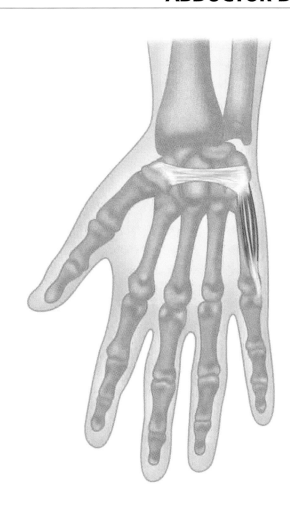

Latin, *abductor*, away from; *digit*, finger; *minimi*, smallest.

This is the most superficial muscle of the hypothenar eminence. The others are: flexor digiti minimi brevis and opponens digiti minimi.

Origin
Pisiform bone. Tendon of flexor carpi ulnaris.

Insertion
Ulna (medial) side of base of proximal phalanx of little finger.

Action
Abducts the little finger. A surprisingly powerful muscle which particularly comes into play when fingers are spread to grasp a large object.

Nerve
Ulnar nerve C(7), **8**, T**1**.

Artery
Deep palmar branches of ulnar artery
(from brachial artery).

Basic functional movement
Example: Holding a large ball.

OPPONENS DIGITI MINIMI

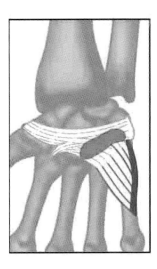

Palmar view, right hand.

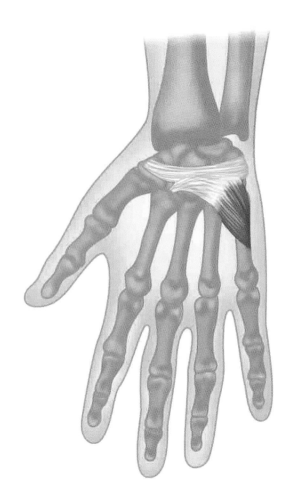

Latin, *opponens*, opposing; *digit*, finger; *minimi*, smallest.

Part of the hypothenar eminence, lying deep to abductor digiti minimi.

Origin
Hook of hamate. Anterior surface of flexor retinaculum.

Insertion
Entire length of medial (ulnar) border of fifth metacarpal.

Action
Pulls metacarpal of the little finger forward and rotates it laterally, so deepening the hollow of the hand, and enabling the pad of the little finger to contact the pad of the thumb.

Nerve
Ulnar nerve C(7), **8**, T**1**.

Artery
Deep palmar branches of ulnar artery
(from brachial artery).

Basic functional movement
Example: Holding a thread within the fingertips (along with the other fingertips).

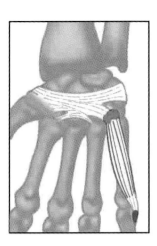

Palmar view, right hand.

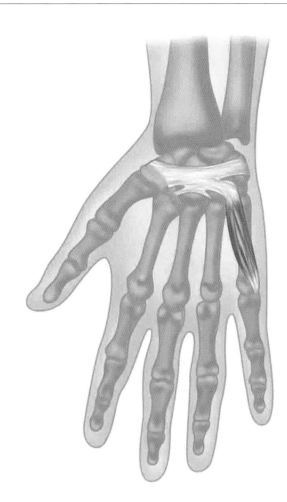

Latin, *flexor*, to flex; *digit*, finger; *minimi*, smallest; *brevis*, short.

Part of the hypothenar eminence. May be absent or fused with a neighbouring muscle.

Origin
Hook of hamate. Anterior surface of flexor retinaculum.

Insertion
Ulna (medial) side of base of proximal phalanx of little finger.

Action
Flexes little finger at the metacarpophalangeal joint.

Nerve
Ulnar nerve C(7), **8**, T**1**.

Artery
Ulnar artery
(from brachial artery).

PALMARIS BREVIS

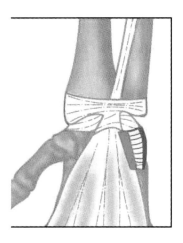

Palmar view, right hand.

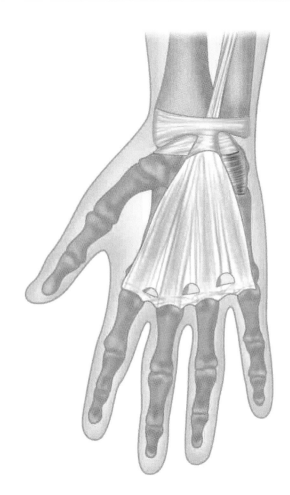

Latin, *palmaris*, palma, palm; *brevis*, short.

This is a small subcutaneous muscle lying over the hypothenar eminence.

Origin
Palmar aponeurosis. Flexor retinaculum.

Insertion
Skin on ulnar border of hand.

Action
Wrinkles the skin on the ulnar border of hand.

Nerve
Ulnar nerve, C(7), **8**, T**1**.

Artery
Ulnar artery
(from brachial artery).

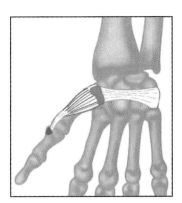

Palmar view, right hand.

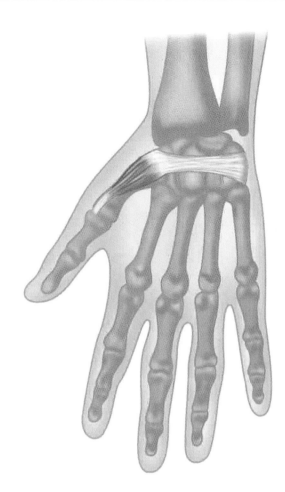

Latin, *abduct*, away from; *pollicis*, of the thumb; *brevis*, short.

This is the most superficial of the muscles of the thenar eminence. The others are flexor pollicis brevis and opponens pollicis.

Origin
Flexor retinaculum. Tubercle of trapezium. Tubercle of scaphoid.

Insertion
Radial side of base of proximal phalanx of thumb.

Action
Abducts thumb and moves it anteriorly (as in typing or playing the piano). Assists in opposition of thumb.

Nerve
Median nerve (C6, 7, 8, T1).

Artery
Superficial palmar branches of radial artery
(from brachial artery).

Basic functional movement
Example: Typing.

OPPONENS POLLICIS

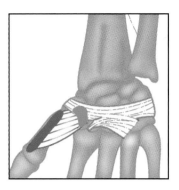

Palmar view, right hand.

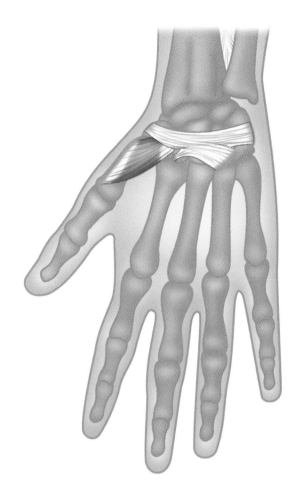

Latin, *opponens*, opposing; *pollicis*, of the thumb.

Part of the thenar eminence, usually partly fused with flexor pollicis brevis and deep to abductor pollicis brevis.

Origin
Flexor retinaculum. Tubercle of trapezium.

Insertion
Entire length of radial border of first metacarpal.

Action
Opposes (i.e. abducts, then slightly medially rotates, followed by flexion and adduction) the thumb so that the pad of the thumb can be drawn into contact with the pads of the fingers.

Nerve
Median nerve (C6, 7, 8, T1)

Artery
Superficial palmar branches of radial artery
(from brachial artery).

Basic functional movement
Example: Picking up small object between thumb and fingers.

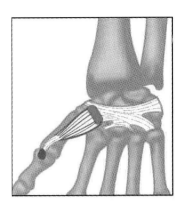

Palmar view, right hand.

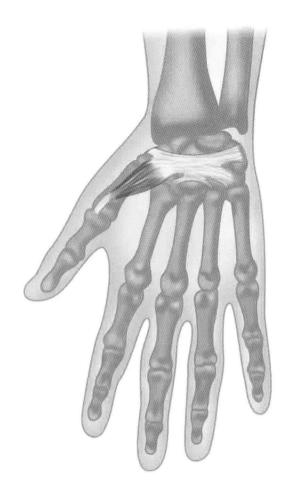

Latin, *flexor*, to flex; *pollicis*, of the thumb; *brevis*, short.

Part of the thenar eminence, together with opponens pollicis, to which it is usually partly fused, and abductor pollicis brevis.

Origin
Superficial head: flexor retinaculum. Trapezium.
Deep head: trapezoid. Capitate.

Insertion
Radial side of base of proximal phalanx of thumb.

Action
Flexes the metacarpophalangeal and carpometacarpal joints of the thumb. Assists in opposition of the thumb towards the little finger.

Nerve
Superficial head: median nerve (C6, 7, 8, T1).
Deep head: ulnar (C**8**, T1).

Artery
Superficial palmar branches of radial artery (from brachial artery).

Basic functional movement
Example: Holding a thread between thumb and fingertips.

ADDUCTOR POLLICIS

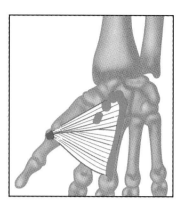

Palmar view, right hand.

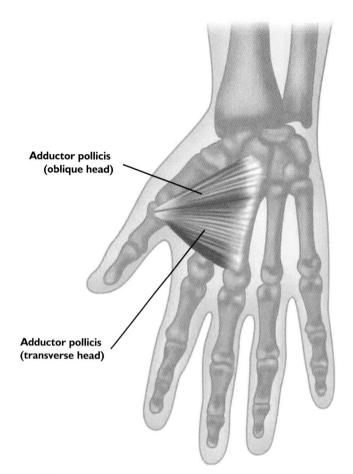

Adductor pollicis
(oblique head)

Adductor pollicis
(transverse head)

Latin, *adduct*, toward; *pollicis*, of the thumb.

Origin
Oblique fibres: anterior surfaces of second and third metacarpals, capitate and trapezoid.
Transverse fibres: palmar surface of third metacarpal bone.

Insertion
Ulna (medial) side of base of proximal phalanx of thumb.

Action
Adducts the thumb.

Nerve
Deep ulnar nerve, C8, T1.

Artery
Palmar metacarpal arteries of deep palmar arterial arch.

Basic functional movement
Example: Gripping a jam jar lid to screw it on.

Muscles of the Hip and Thigh

13

MUSCLES OF THE BUTTOCK

The bulk of the buttock is mainly formed by the gluteus maximus, which is the largest and most superficial muscle of the group, lying posterior to smaller muscles such as gluteus medius and gluteus minimus. Tensor fasciae latae is included as the most anterior muscle of the group. Other muscles, such as the gemelli, quadratus femoris, obturator internus and piriformis are sometimes grouped with muscles of the buttock, but have here been dealt with under muscles of the hip.

Gluteus maximus

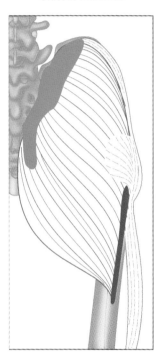

Tensor fasciae latae

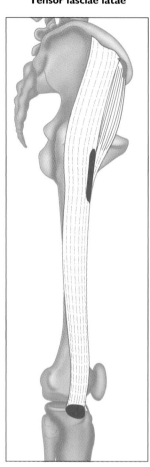

Gluteus medius

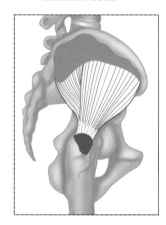

Gluteus minimus

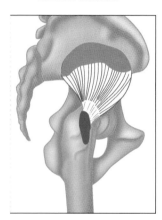

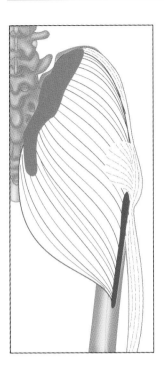

Posterior view, right leg.

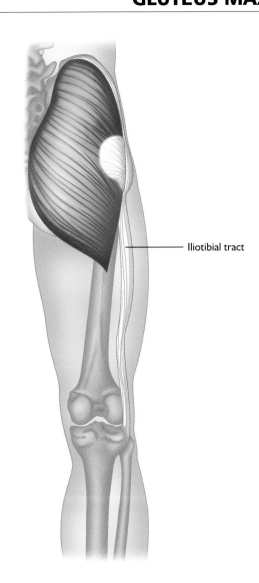

Iliotibial tract

Greek, *gloutos*, buttock; *maximus*, biggest.

The gluteus maximus is the most coarsely fibred and heaviest muscle in the body.

Origin
Outer surface of ilium behind posterior gluteal line and portion of bone superior and posterior to it. Adjacent posterior surface of sacrum and coccyx. Sacrotuberous ligament. Aponeurosis of erector spinae.

Insertion
Deep fibres of distal portion: gluteal tuberosity of femur.
Remaining fibres: iliotibial tract of fascia lata.

Action
Upper fibres: laterally rotate hip joint. May assist in abduction of hip joint.

Lower fibres: extend and laterally rotate hip joint (forceful extension as in running or rising from sitting). Extend trunk. Assists in adduction of hip joint.
Through its insertion into the iliotibial tract, helps to stabilize the knee in extension.

Nerve
Inferior gluteal nerve, L5, S1, 2.

Artery
Inferior and superior gluteal arteries
via internal iliac artery (a branch of the common iliac artery from abdominal aorta).
First perforating branch of the deep femoral artery (via external iliac artery).

Basic functional movement
Examples: Walking upstairs. Rising from sitting.

TENSOR FASCIAE LATAE

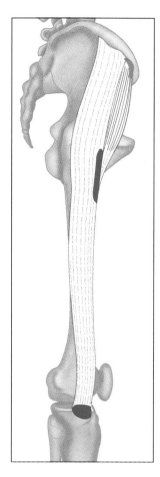

Lateral view, right leg.

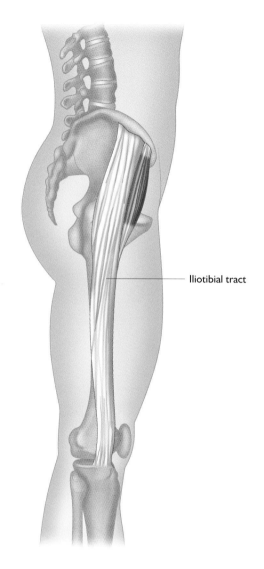

Iliotibial tract

Latin, *tensor*, stretcher, puller; *fascia(e)*, band(s); *latae*, broad.

This muscle lies anterior to gluteus maximus, on the lateral side of the hip.

Origin
Anterior part of outer lip of iliac crest, and outer surface of anterior superior iliac spine.

Insertion
Joins iliotibial tract just below level of greater trochanter.

Action
Flexes, abducts and medially rotates the hip joint. Tenses the fascia lata, thus stabilizing the knee. Redirects the rotational forces produced by gluteus maximus.

Nerve
Superior gluteal nerve, L4, 5, S1.

Artery
Superior gluteal artery
via internal iliac artery (a branch of the common iliac artery from abdominal aorta).
Lateral circumflex femoral artery
via deep femoral artery (from external iliac artery).

Basic functional movement
Example: Walking.

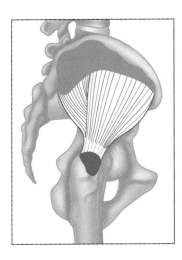

Lateral view, right leg.

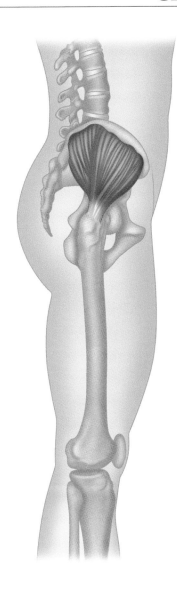

Greek, *gloutos*, buttock; *medius*, middle.

This muscle is mostly deep to and therefore obscured by gluteus maximus, but appears on the surface between gluteus maximus and tensor fasciae latae. During walking, this muscle, with gluteus minimus, prevents the pelvis from dropping towards the non weight-bearing leg.

Origin
Outer surface of ilium inferior to iliac crest, between the posterior gluteal line and the anterior gluteal line.

Insertion
Oblique ridge on lateral surface of greater trochanter of femur.

Action
Abducts the hip joint. Anterior fibres medially rotate and may assist in flexion of the hip joint. Posterior fibres slightly laterally rotate the hip joint.

Nerve
Superior gluteal nerve, L4, **5**, S1.

Artery
Superior gluteal artery
via internal iliac artery (a branch of the common iliac artery from abdominal aorta).

Basic functional movement
Example: Stepping sideways over an object such as a low fence.

GLUTEUS MINIMUS

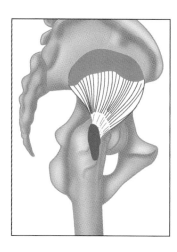

Lateral view, right leg.

Greek, *gloutos*, buttock; *minimus*, smallest.

This muscle is situated antero-inferior and deep to gluteus medius, whose fibres obscure it.

Origin
Outer surface of ilium between anterior and inferior gluteal lines.

Insertion
Anterior border of greater trochanter.

Action
Abducts, medially rotates, and may assist in flexion of the hip joint.

Nerve
Superior gluteal nerve, L4, **5**, S1.

Artery
Superior gluteal artery
via internal iliac artery (a branch of the common iliac artery from abdominal aorta).

Basic functional movement
Example: Stepping sideways over an object such as a low fence.

The muscles of the hip are relatively small muscles originating from the sacrum and / or inner surface of the pelvis, to insert on or near the greater trochanter of femur. They all have a role in lateral rotation of the hip joint. They also, especially piriformis and obturator internus, help hold the head of the femur in the acetabulum, similar to the role of the rotator cuff muscles of the shoulder joint.

Piriformis

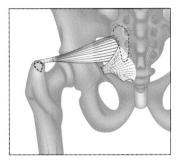

Gemellus superior

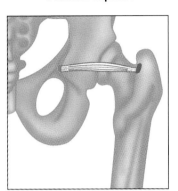

Obturator externus

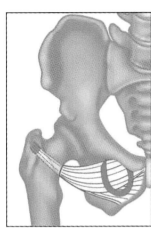

Obturator internus

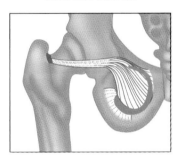

Gemellus inferior

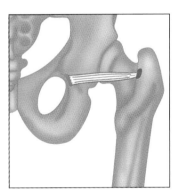

Quadratus femoris

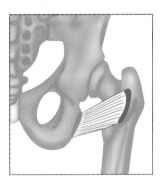

PIRIFORMIS

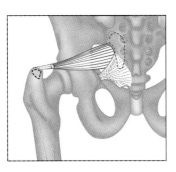

Posterior view.
Origin on anterior of sacrum.

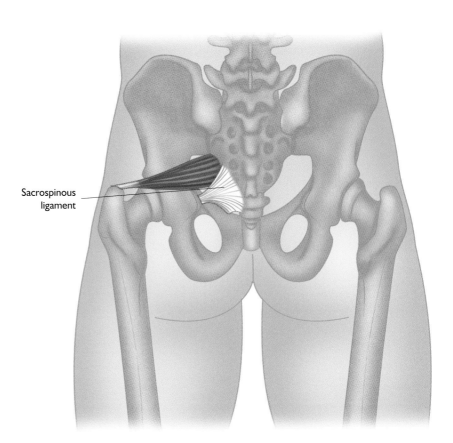

Sacrospinous ligament

Latin, *pirum*, a pear, *piriform*, pear-shaped.

Piriformis leaves the pelvis by passing through the greater sciatic foramen.

Origin
Internal surface of sacrum. Sacrotuberous ligament.

Insertion
Superior border of greater trochanter of femur.

Action
Laterally rotates hip joint. Abducts the thigh when hip is flexed. Helps hold head of femur in acetabulum.

Nerve
Ventral rami of lumbar nerve, L(5) and sacral nerves, S**1, 2.**

Artery
Superior and inferior gluteal arteries
via internal iliac artery (a branch of the common iliac artery from abdominal aorta).

Basic functional movement
Example: Taking first leg out of car.

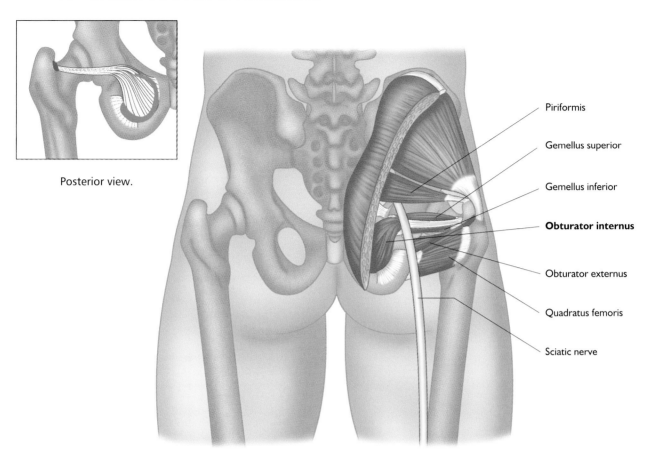

Posterior view.

Piriformis

Gemellus superior

Gemellus inferior

Obturator internus

Obturator externus

Quadratus femoris

Sciatic nerve

Posterior view.

Latin, *obturator*, obstructor; *internus*, internal.

The obturator internus is very closely associated with the two gemelli, in relation to both action and position. It leaves the pelvis by passing through the lesser sciatic foramen.

Origin
Inner surface of obturator membrane and margin of obturator foramen. Inner surface of ischium, pubis and ilium.

Insertion
Medial surface of greater trochanter of femur above trochanteric fossa.

Action
Laterally rotates hip joint. Abducts the thigh when the hip is flexed. Helps hold head of femur in acetabulum.

Nerve
Nerve to obturator internus, a branch of the ventral rami of lumbar nerve, L**5** and sacral nerves, S**1**, **2**.

Artery
Superior and inferior gluteal arteries and obturator artery
via internal iliac artery (a branch of the common iliac artery from abdominal aorta).

Basic functional movement
Example: Taking first leg out of car.

GEMELLUS SUPERIOR / GEMELLUS INFERIOR

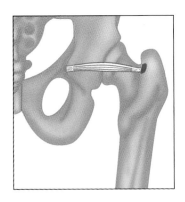

Posterior view.

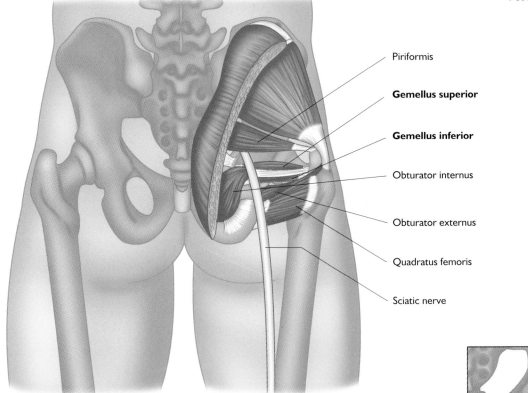

Piriformis

Gemellus superior

Gemellus inferior

Obturator internus

Obturator externus

Quadratus femoris

Sciatic nerve

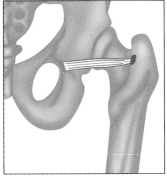

Posterior view.

Latin, *gemellus*, a twin; *superior*, above.

Both gemelli are accessory to the obturator internus, providing additional origins from the margins of the lesser sciatic notch.

Origin
External surface of ischial spine.

Insertion
With tendon of obturator internus into medial surface of greater trochanter of femur.

Action
Assists obturator internus to laterally rotate hip joint and abduct the thigh when the hip is flexed.

Nerve
Nerve to obturator internus, a branch of the ventral rami of lumbar nerve, L5 and sacral nerves S1, 2.

Artery
Inferior gluteal artery
via internal iliac artery (a branch of the common iliac artery from abdominal aorta).

Basic functional movement
Example: Taking first leg out of car.

Latin, *gemellus*, a twin; *inferior*, below.

Origin
Upper margin of ischial tuberosity.

Insertion
With tendon of obturator internus into medial surface of greater trochanter of femur.

Action
Assists obturator internus to laterally rotate hip joint and abduct the thigh when the hip is flexed.

Nerve
Branch of nerve to quadratus femoris, a branch of lumbosacral plexus, L4, 5, S1, (2).

Artery
Inferior gluteal artery
via internal iliac artery (a branch of the common iliac artery from abdominal aorta).

Basic functional movement
Example: Taking first leg out of car.

OBTURATOR EXTERNUS / QUADRATUS FEMORIS

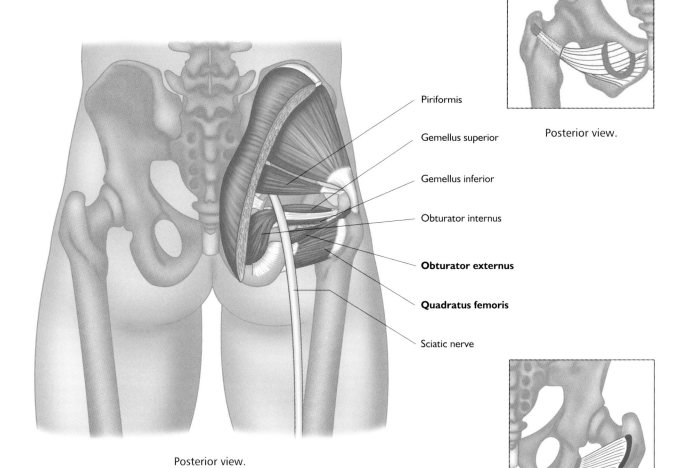

Piriformis

Gemellus superior

Gemellus inferior

Obturator internus

Obturator externus

Quadratus femoris

Sciatic nerve

Posterior view.

Posterior view.

Posterior view.

Latin, *obturator*, obstructor; *externus*, external.

This muscle is often grouped with the hip adductors, but is placed in this section because of its similarity and proximity to the other short lateral rotators of the hip.

Origin
Rami of pubis and ischium. External surface of obturator membrane.

Insertion
Trochanteric fossa of femur.

Action
Laterally rotates hip joint. May assist in adduction of hip joint.

Nerve
Posterior division of the obturator nerve, L**3**, **4**.

Artery
Obturator artery
via internal iliac artery (a branch of the common iliac artery from abdominal aorta), plus can also be supplied by medial circumflex arteries (from deep femoral artery).

Basic functional movement
Example: Clicking heels together 'military style'.

Latin, *quadratus*, squared; *femoris*, of the thigh.

This muscle is often fused with either or both the gemellus inferior, which lies above, and the upper fibres of adductor magnus, which lies below.

Origin
Lateral border of ischial tuberosity.

Insertion
Quadrate line that extends distally below intertrochanteric crest.

Action
Laterally rotates hip joint.

Nerve
Nerve to quadratus femoris, a branch of lumbosacral plexus, L**4**, **5**, S**1**, (2).
This nerve also supplies the gemellus inferior.

Artery
Inferior gluteal artery
via internal iliac artery (a branch of the common iliac artery from abdominal aorta), plus can also be supplied by medial circumflex arteries (from deep femoral artery).

Basic functional movement
Example: Taking first leg out of car.

MUSCLES OF THE THIGH

The muscles of the thigh consist of three broad groups. The posterior thigh consists of the **hamstring group**, namely: *semitendinosus, semimembranosus,* and *biceps femoris*. The hamstrings correspond to the flexors of the elbow in the upper limb. The medial thigh consists of the **adductor group**, which include *adductors magnus, brevis, longus* and *pectineus*. The adductor group corresponds to coracobrachialis in the upper limb. Obturator externus can also be placed in this group, but has been included on page 322 under Muscles of the Hip. The **anterior group** consists of *sartorius* and the four muscles of the *quadriceps femoris*. This group corresponds to the triceps brachii of the upper limb.

Semitendinosus

Biceps femoris

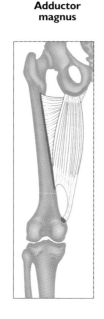

Adductor longus

Pectineus

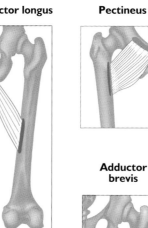

Adductor brevis

Rectus femoris

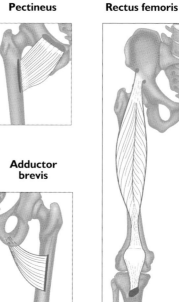

Vastus lateralis

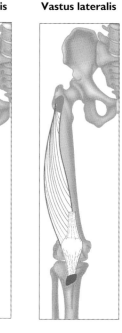

Semimembranosus

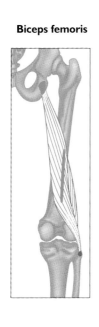

Adductor magnus

Gracilis

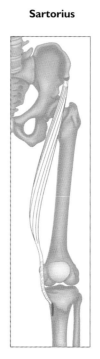

Sartorius

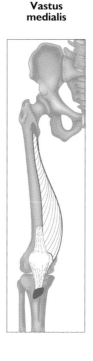

Vastus medialis

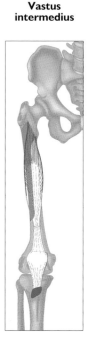

Vastus intermedius

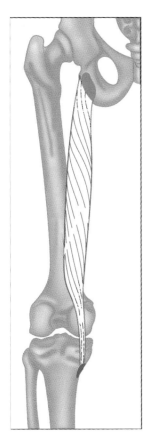

Posterior view.

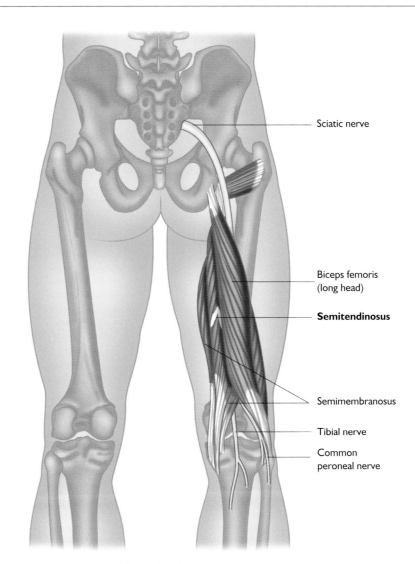

Sciatic nerve

Biceps femoris
(long head)

Semitendinosus

Semimembranosus

Tibial nerve

Common
peroneal nerve

Posterior view.

Latin, *semi*, half; *tendinosus*, tendinous.

Semitendinosus is the central part of the hamstring group. During running, the hamstrings slow down the leg at the end of its forward swing and prevent the trunk from flexing at the hip joint.

Origin
Ischial tuberosity.

Insertion
Upper medial surface of shaft of tibia.

Action
Flexes and slightly medially rotates knee joint after flexion. Extends the hip joint.

Nerve
Two branches from the tibial part of sciatic nerve, L4, 5, S1, 2.

Artery
Perforating branches of the deep femoral artery
via femoral artery (a continuation of the external iliac artery).
Inferior gluteal artery via internal iliac artery
(a branch of the common iliac artery from abdominal aorta).

Basic functional movement
During running, the hamstrings slow down the leg at the end of its forward swing and prevent the trunk from flexing at the hip joint.

SEMIMEMBRANOSUS / BICEPS FEMORIS

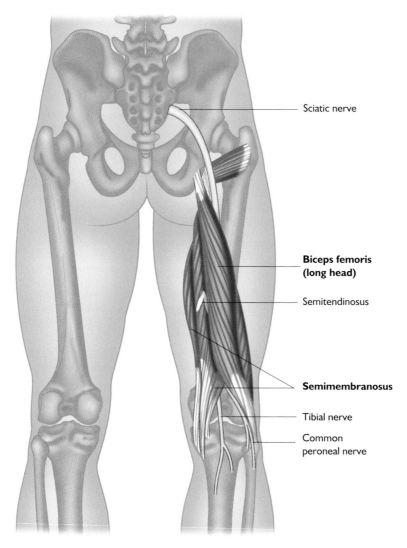

Sciatic nerve

Biceps femoris (long head)

Semitendinosus

Semimembranosus

Tibial nerve

Common peroneal nerve

Posterior view.

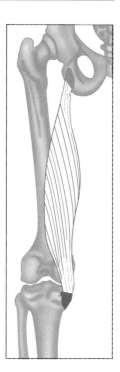

Posterior view.

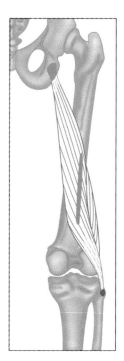

Posterior view.

Latin, *semi*, half; *membranosus*, membranous.

Medial part of hamstring group. Most of its belly is deep to semitendinosus and the long head of biceps femoris.

Origin
Ischial tuberosity.

Insertion
Posteromedial surface of medial condyle of tibia.

Action
Flexes and slightly medially rotates knee joint after flexion. Extends the hip joint.

Nerve
Two branches from the tibial part of sciatic nerve, L4, **5**, S**1**, 2.

Artery
Perforating branches of the deep femoral artery via femoral artery (a continuation of the external iliac artery).
Inferior gluteal artery via internal iliac artery (a branch of the common iliac artery from abdominal aorta).

Basic functional movement
During running, the hamstrings slow down the leg at the end of its forward swing and prevent the trunk from flexing at the hip joint.

Latin, *biceps*, two headed muscle; *femoris*, of the thigh.

Lateral part of hamstring group.

Origin
Long head: ischial tuberosity. Sacrotuberous ligament.
Short head: linea aspera. Upper two-thirds of supracondylar line. Lateral intermuscular septum.

Insertion
Lateral side of head of fibula. Lateral condyle of tibia.

Action
Both heads flex the knee joint (and laterally the flexed knee joint).
The long head also extends the hip joint.

Nerve
Long head: tibial portion of sciatic nerve, L5, S**1**, **2**, 3.
Short head: common fibular (fibular/peroneal) part of sciatic nerve, L5, S**1**, **2**.

Artery
Perforating branches of the deep femoral artery via femoral artery (a continuation of the external iliac artery).
Inferior gluteal artery via internal iliac artery (a branch of the common iliac artery from abdominal aorta).

Basic functional movement
During running, the hamstrings slow down the leg at the end of its forward swing and prevent the trunk from flexing at the hip joint.

ADDUCTOR MAGNUS / ADDUCTOR BREVIS

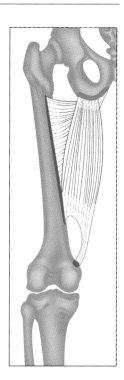

Anterior view.

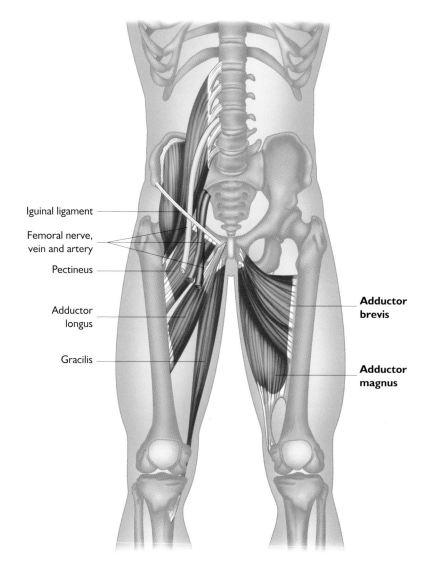

Iguinal ligament

Femoral nerve,
vein and artery

Pectineus

Adductor
longus

Gracilis

**Adductor
brevis**

**Adductor
magnus**

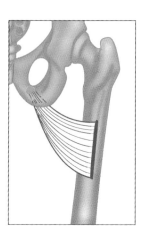

Anterior view.

Latin, *adductor*, toward; *magnus*, large.

The adductor magnus is the largest of the adductor muscle group. Upper fibres are often fused with those of quadratus femoris. The vertical fibres of the ischial part belong morphologically to the hamstring group, and are therefore supplied by the tibial nerve.

Origin
Inferior ramus of pubis. Ramus of ischium (anterior fibres). Ischial tuberosity (posterior fibres).

Insertion
Whole length of femur, along linea aspera and medial supracondylar line to adductor tubercle on medial epicondyle of femur.

Action
Upper fibres adduct and laterally rotate hip joint. Vertical fibres from ischium may assist in weak extension of the hip joint.

Nerve
Posterior division of obturator nerve, L2, **3**, **4**. Tibial portion of sciatic nerve, L4, 5, S1.

Artery
Obturator artery
via internal iliac artery (a branch of the common iliac artery from abdominal aorta).
Medial circumflex femoral artery
(from deep femoral artery).

Basic functional movement
Example: Bringing second leg in or out of car.

Latin, *adductor*, toward; *brevis*, short.

The adductor brevis lies anterior to adductor magnus.

Origin
Outer surface of inferior ramus of pubis.

Insertion
Lower two-thirds of pectineal line and upper half of linea aspera.

Action
Adducts hip joint. Flexes extended femur at hip joint. Extends flexed femur at hip joint. Assists lateral rotation of hip joint.

Nerve
Anterior division of obturator nerve (L2–L4). Sometimes the posterior division also supplies a branch to it.

Artery
Obturator artery
via internal iliac artery (a branch of the common iliac artery from abdominal aorta).
Deep femoral artery
Medial circumflex femoral artery
(from deep femoral artery).

Basic functional movement
Example: Bringing second leg in or out of car.

ADDUCTOR LONGUS

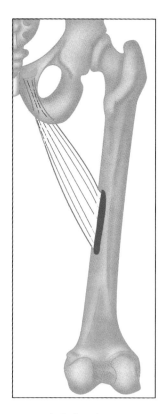

Anterior view.

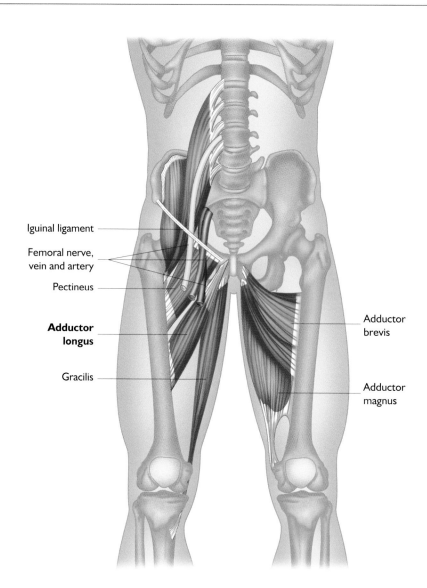

Iguinal ligament

Femoral nerve,
vein and artery

Pectineus

**Adductor
longus**

Gracilis

Adductor
brevis

Adductor
magnus

Latin, *adductor*, toward; *longus*, long.

This is the most anterior of the three adductor muscles. The lateral border of the upper fibres of adductor longus form the medial border of the **femoral triangle** (sartorius forms the lateral boundary; the inguinal ligament forms the superior boundary).

Origin
Anterior surface of pubis at junction of crest and symphysis.

Insertion
Middle third of medial lip of linea aspera.

Action
Adducts hip joint. Flexes extended femur at hip joint. Extends flexed femur at hip joint. Assists lateral rotation of hip joint.

Nerve
Anterior division of obturator nerve, L**2**, **3**, 4.

Artery
Obturator artery
via internal iliac artery (a branch of the common iliac artery from abdominal aorta).
Deep femoral artery
Medial circumflex femoral artery
(from deep femoral artery).

Basic functional movement
Example: Bringing second leg in or out of car.

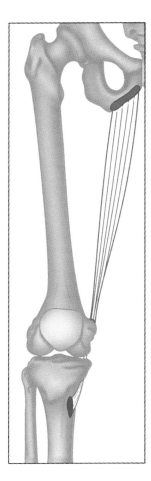

Anterior view, right leg.

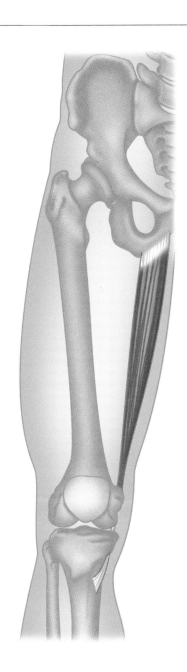

Latin, slender, delicate.

Gracilis descends down the medial side of the thigh anterior to semimembranosus.

Origin
Lower half of symphysis pubis and inferior ramus of pubis.

Insertion
Upper part of medial surface of shaft of tibia.

Action
Adducts hip joint. Flexes knee joint. Medially rotates knee joint when flexed.

Nerve
Anterior division of obturator nerve, L**2**, **3**, **4**.

Artery
Obturator artery
via internal iliac artery (a branch of the common iliac artery from abdominal aorta),
plus can be supplied by medial circumflex femoral artery (from deep femoral artery).

Basic functional movement
Example: Sitting with knees pressed together.

PECTINEUS

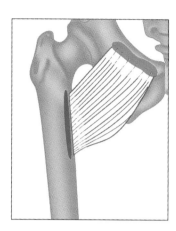

Anterior view, right leg.

Latin, *pecten*, comb, *pectenate*, shaped like a comb.

Pectineus is sandwiched between the psoas major and adductor longus.

Origin
Pecten of pubis, between iliopubic (iliopectineal) eminence and pubic tubercle.

Insertion
Pectineal line, from lesser trochanter to linea aspera of femur.

Action
Adducts the hip joint. Flexes the hip joint.

Nerve
Femoral nerve, L**2**, **3**, 4. Occasionally receives an additional branch from the obturator nerve L3.

Artery
Medial circumflex femoral artery (from deep femoral artery).

Basic functional movement
Example: Walking along a straight line.

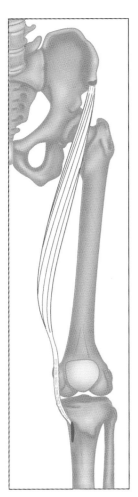

Anterior view.

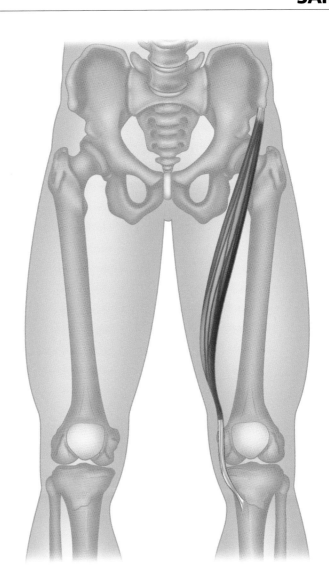

Latin, tailor.

Sartorius is the most superficial muscle of the anterior thigh. It is also the longest strap muscle in the body. The medial border of the upper third of this muscle forms the lateral boundary of the femoral triangle (adductor longus forms the medial boundary; the inguinal ligament forms the superior boundary). The action of sartorius is to put the lower limbs in the cross-legged seated position of the tailor (hence its name from the Latin).

Origin
Anterior superior iliac spine and area immediately below it.

Insertion
Upper part of medial surface of tibia, near anterior border.

Action
Flexes hip joint (helping to bring leg forward in walking or running). Laterally rotates and abducts the hip joint. Flexes knee joint. Assists in medial rotation of the tibia on the femur after flexion. These actions may be summarized by saying that it places the heel on the knee of the opposite limb.

Nerve
Two branches from the femoral nerve, L**2**, **3**, (4).

Artery
Lateral femoral circumflex artery (from deep femoral artery).
Saphenous branch of the descending genicular artery (from femoral artery).

Basic functional movement
Example: Sitting cross-legged.

RECTUS FEMORIS / VASTUS LATERALIS

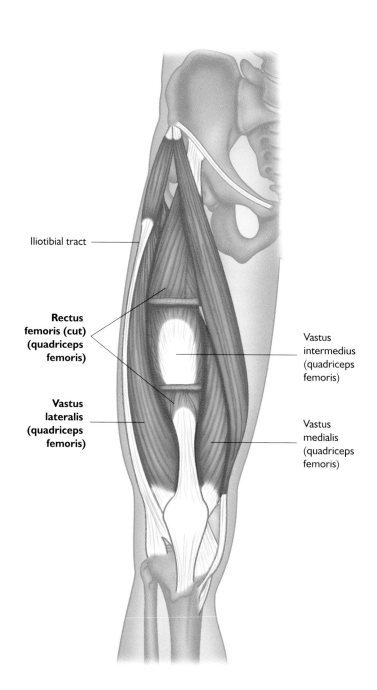

Iliotibial tract

Rectus femoris (cut) (quadriceps femoris)

Vastus lateralis (quadriceps femoris)

Vastus intermedius (quadriceps femoris)

Vastus medialis (quadriceps femoris)

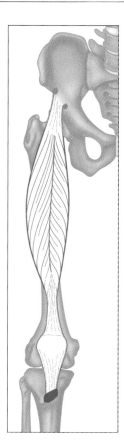

Anterior view, right leg.

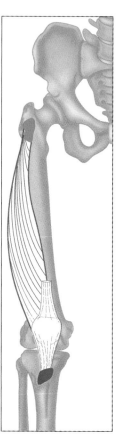

Anterior view, right leg.

Latin, *rectum*, straight; *femoris*, of the thigh.

Rectus femoris is part of the quadriceps femoris. It has two heads of origin. The reflected head is in the line of pull of the muscle in four-footed animals, whereas the straight head seems to have developed in humans as a result of the upright posture. It is a spindle shaped bipennate muscle.

Origin
Straight head (anterior head): anterior inferior iliac spine.
Reflected head (posterior head): groove above acetabulum (on ilium).

Insertion
Patella, then via patellar ligament to tuberosity of tibia.

Action
Extends the knee joint and flexes the hip joint (particularly in combination, as in kicking a ball). Assists iliopsoas to flex the trunk on the thigh. Prevents flexion at knee joint as heel strikes the ground during walking.

Nerve
Femoral nerve, L2, 3, 4.

Artery
Lateral femoral circumflex artery
(from deep femoral artery).

Basic functional movement
Examples: Walking up stairs. Cycling.

Latin, *vastus*, great or vast; *lateral*, to the side.

Part of the quadriceps femoris. The quadriceps straighten the knee when rising from sitting, during walking and climbing. The vasti muscles as a group pay out to control the movement of sitting down.

Origin
Proximal part of intertrochanteric line. Anterior and inferior borders of greater trochanter. Gluteal tuberosity. Upper half of lateral lip of linea aspera of femur.

Insertion
Lateral margin of patella, then via patellar ligament to tuberosity of tibia.

Action
Extends the knee joint. Prevents flexion at knee joint as heel strikes the ground during walking.

Nerve
Femoral nerve, L2, 3, 4.

Artery
Lateral femoral circumflex artery
(from deep femoral artery), plus can be supplied by perforating branches of the deep femoral artery.

Basic functional movement
Examples: Walking up stairs. Cycling.

VASTUS MEDIALIS / VASTUS INTERMEDIUS

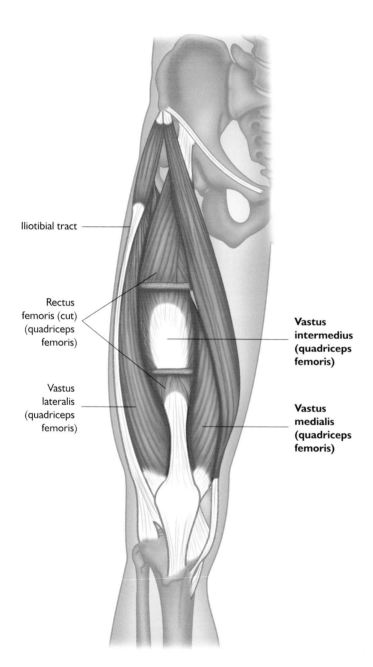

Iliotibial tract

Rectus
femoris (cut)
(quadriceps
femoris)

Vastus
lateralis
(quadriceps
femoris)

**Vastus
intermedius
(quadriceps
femoris)**

**Vastus
medialis
(quadriceps
femoris)**

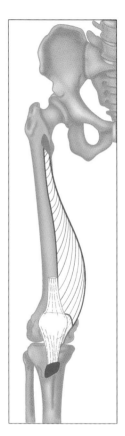

Anterior view, right leg.

Anterior view, right leg.

Latin, *vastus*, great or vast; *medial*, middle.

Part of the quadriceps femoris. Vastus medialis is larger and heavier than vastus lateralis.

Origin
Distal one-half of intertrochanteric line. Medial lip of linea aspera. Medial supracondylar line. Medial intermuscular septum.

Insertion
Medial margin of patella, then via patellar ligament to tuberosity of tibia. Medial condyle of tibia.

Action
Extends knee joint. Prevents flexion at knee joint as heel strikes the ground during walking.

Nerve
Femoral nerve, L2, **3**, **4**.

Artery
Lateral femoral circumflex artery
(from deep femoral artery).
Saphenous branch of the descending genicular artery
(from femoral artery).
Insertion also supplied by the medial superior genicular branch of the popliteal artery (a continuation of the femoral artery).

Basic functional movement
Examples: Walking up stairs. Cycling.

Latin, *vastus*, great or vast; *intermedial*, between the middle.

Vastus intermedius is the deepest part of the quadriceps femoris. This muscle has a membranous tendon on its anterior surface, to allow a gliding movement between itself and the rectus femoris that overlies it.

Origin
Anterior and lateral surfaces of upper two-thirds of shaft of femur. Lower half of linea aspera. Lateral intermuscular septum. Upper part of lateral supracondylar line.

Insertion
Deep surface of quadriceps tendon, then via patellar ligament to tuberosity of tibia.

Action
Extends knee joint. Prevents flexion at knee joint as heel strikes the ground during walking.

Nerve
Femoral nerve, L2, **3**, **4**.

Artery
Lateral femoral circumflex artery
(from deep femoral artery).

Basic functional movement
Examples: Walking up stairs. Cycling.

Muscles of the
Leg and Foot

14

MUSCLES OF THE LEG

The leg comprises three muscle groups. 1) The extensors (dorsiflexors) within the **anterior compartment**: *tibialis anterior, extensor hallucis longus, extensor digitorum longus*, and *peroneus tertius*. 2) The **peroneal compartment** on the lateral side: *peroneus longus* and *brevis*. 3) The flexor (plantar flexor) muscles within the **posterior compartment**: *gastrocnemius, soleus, plantaris, popliteus, flexor digitorum longus, flexor hallucis longus*, and *tibialis posterior*.

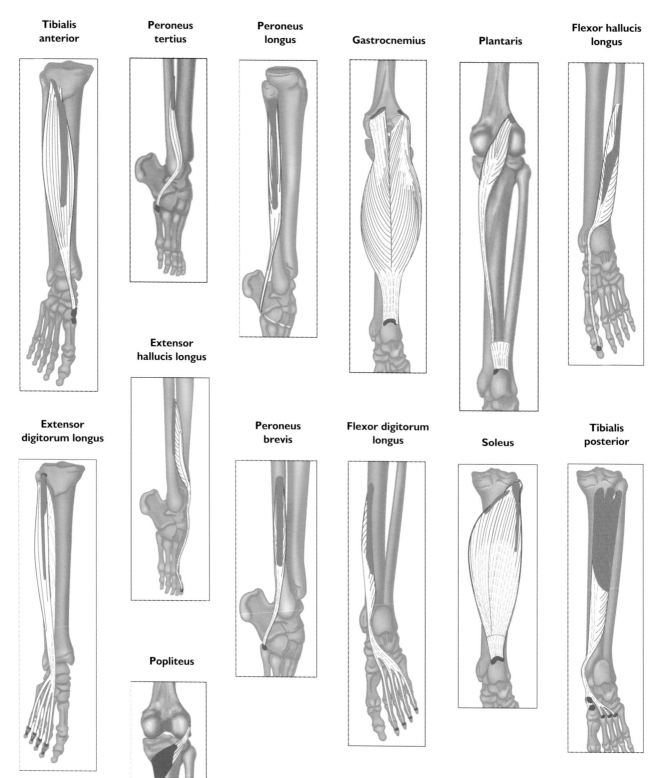

Tibialis anterior

Peroneus tertius

Peroneus longus

Gastrocnemius

Plantaris

Flexor hallucis longus

Extensor hallucis longus

Extensor digitorum longus

Peroneus brevis

Flexor digitorum longus

Soleus

Tibialis posterior

Popliteus

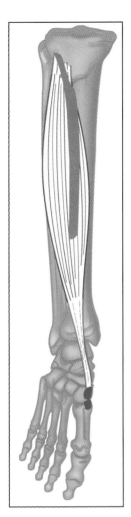

Anterior view, right leg.

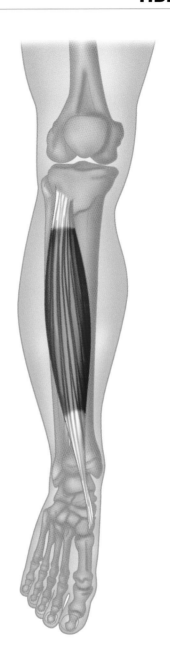

Latin, *tibia*, pipe or flute / shinbone; *anterior*, before.

Origin
Lateral condyle of tibia. Upper half of lateral surface of tibia. Interosseous membrane.

Insertion
Medial and plantar surface of medial cuneiform bone. Base of first metatarsal.

Action
Dorsiflexes the ankle joint. Inverts the foot.

Nerve
Deep peroneal nerve, L4, **5**, S1.

Artery
Anterior tibial artery
(from popliteal artery, a continuation of the femoral artery).

Basic functional movement
Example: Walking and running (helps prevent the foot from slapping onto the ground after the heel strikes. Lifts the foot clear of the ground as the leg swings forward).

EXTENSOR DIGITORUM LONGUS

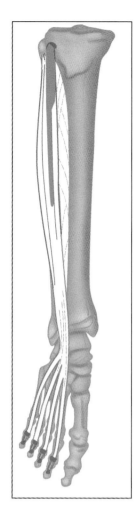

Anterior view, right leg.

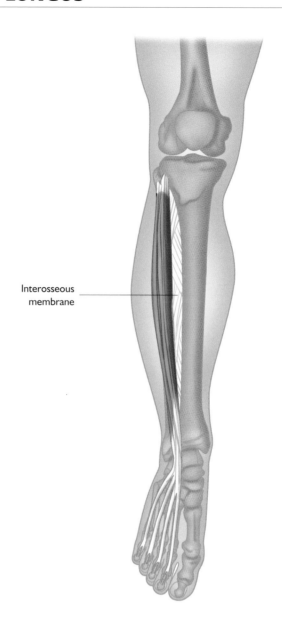

Interosseous
membrane

Latin, *extensor*, to extend; *digit*, toe; *longus*, long.

Like the corresponding tendons in the hand, this muscle forms extensor hoods on the dorsum of the proximal phalanges of the foot. These hoods are joined by the tendons of the lumbricales and extensor digitorum brevis, but not by the interossei.

Origin
Lateral condyle of tibia. Upper two-thirds of anterior surface of fibula. Upper part of interosseous membrane.

Insertion
Along dorsal surface of the four lateral toes. Each tendon dividing to attach to the bases of the middle and distal phalanges.

Action
Extends toes at the metatarsophalangeal joints. Assists the extension of the interphalangeal joints. Assists in dorsiflexion of ankle joint and eversion of the foot.

Nerve
Fibular (peroneal) nerve, L4, **5**, S1.

Artery
Anterior tibial artery
(from popliteal artery, a continuation of the femoral artery).

Basic functional movement
Example: Walking up the stairs (ensuring the toes clear the steps).

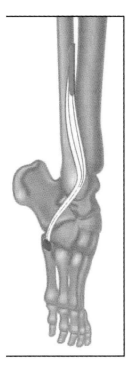

Lateral view, right leg.

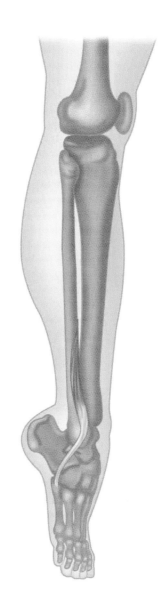

Latin, *fibula*, pin / buckle; *tertius*, third.

This muscle is a partially separated lower lateral part of extensor digitorum longus.

Origin
Lower third of anterior surface of fibula and interosseous membrane.

Insertion
Dorsal surface of base of fifth metatarsal.

Action
Dorsiflexes ankle joint. Everts foot.

Nerve
Deep fibular (peroneal) nerve, L4, 5, S1.

Artery
Anterior tibial artery
(from popliteal artery, a continuation of the femoral artery).

Basic functional movement
Examples: Walking and running.

EXTENSOR HALLUCIS LONGUS

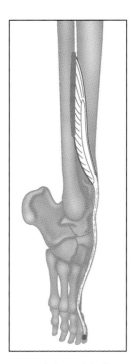

Lateral view, right leg.

Latin, *extensor*, to extend; *hallux*, big toe; *longus*, long.

This muscle lies between and deep to tibialis anterior and extensor digitorum longus.

Origin
Middle half of anterior surface of fibula and adjacent interosseous membrane.

Insertion
Base of distal phalanx of great toe.

Action
Extends all the joints of the big toe. Dorsiflexes the ankle joint. Assists in inversion of the foot.

Nerve
Deep fibular (peroneal) nerve, L4, 5, S1.

Artery
Anterior tibial artery
(from popliteal artery, a continuation of the femoral artery).

Basic functional movement
Example: Walking up the stairs (ensuring the big toe clears the steps).

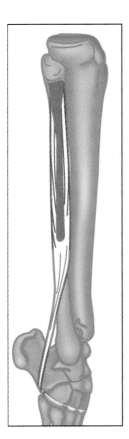

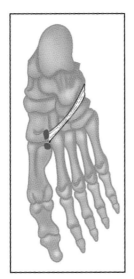

Plantar view, right leg.

Lateral view, right leg.

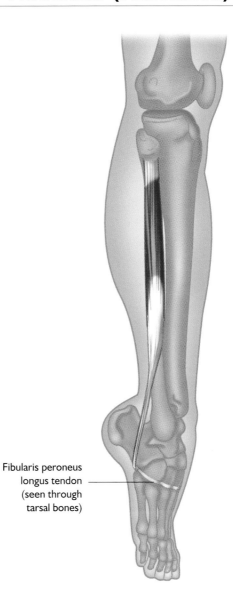

Fibularis peroneus
longus tendon
(seen through
tarsal bones)

Latin, *fibula*, pin / buckle; *longus*, long.

The course of the tendon of insertion of fibularis longus helps maintain the transverse and lateral longitudinal arches of the foot.

Origin
Upper two-thirds of lateral surface of fibula. Lateral condyle of tibia.

Insertion
Lateral side of medial cuneiform. Base of first metatarsal.

Action
Everts the foot. Assists plantar flexion of ankle joint.

Nerve
Superficial fibular (peroneal) nerve, L4, **5**, S1.

Artery
Peroneal (fibular) artery
via posterior tibial artery (from popliteal artery).

Basic functional movement
Example: Walking on uneven surfaces.

FIBULARIS (PERONEUS) BREVIS

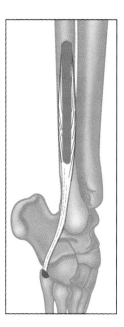

Lateral view, right leg.

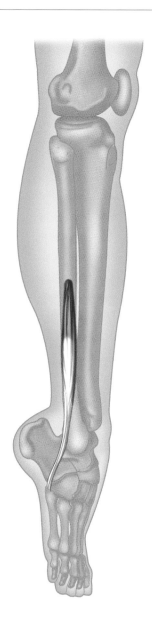

Latin, *fibula*, pin / buckle; *brevis*, short.

A slip of muscle from fibularis brevis often joins the long extensor tendon of the little toe, whereupon it is known as *peroneus digiti minimi*.

Origin
Lower two-thirds of lateral surface of fibula. Adjacent intermuscular septa.

Insertion
Lateral side of base of fifth metatarsal.

Action
Everts ankle joint. Assists plantar flexion of ankle joint.

Nerve
Superficial fibular (peroneal) nerve, L4, **5**, S1.

Artery
Peroneal (fibular) artery
via posterior tibial artery (from popliteal artery).

Basic functional movement
Example: Walking on uneven ground.

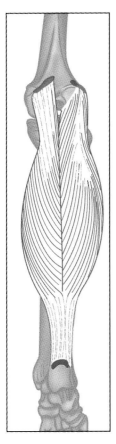

Posterior view,
right leg.

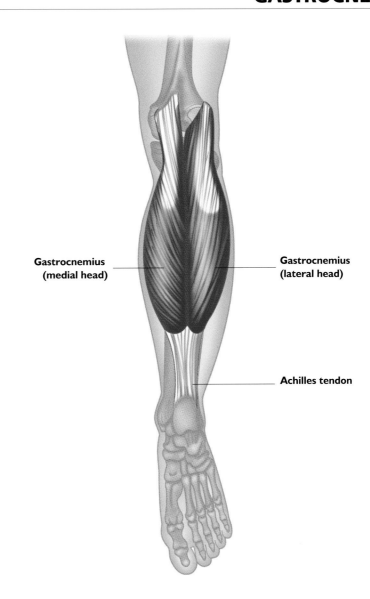

Gastrocnemius
(medial head)

Gastrocnemius
(lateral head)

Achilles tendon

Greek, *gaster*, stomach; *kneme*, leg.

Gastrocnemius is part of the composite muscle known as *triceps surae*, which forms the prominent contour of the calf. The triceps surae comprises: *gastrocnemius, soleus* and *plantaris*. The *popliteal fossa* at the back of the knee is formed inferiorly by the bellies of gastrocnemius and plantaris, laterally by the tendon of biceps femoris, and medially by the tendons of semimembranosus and semitendinosus.

Origin
Medial head: popliteal surface of femur above medial condyle.
Lateral head: lateral condyle and posterior surface of femur.

Insertion
Posterior surface of calcaneus (via the tendo calcaneus; a fusion of the tendons of gastrocnemius and soleus).

Action
Plantar flexes foot at ankle joint. Assists in flexion of knee joint. It is a main propelling force in walking and running.

Nerve
Tibial nerve, S**1, 2**.

Artery
Sural branches of the popliteal artery
(a continuation of the femoral artery).
Posterior tibial artery (from popliteal artery).

Basic functional movement
Example: Standing on tip-toes.

PLANTARIS

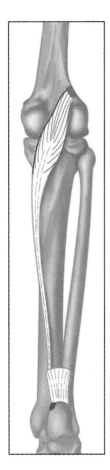

Posterior view,
right leg.

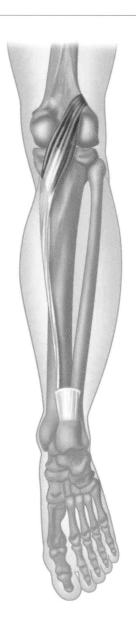

Latin, *planta*, sole of the foot.

Part of the triceps surae. Its long slender tendon is equivalent to the tendon of palmaris longus in the arm.

Origin
Lower part of lateral supracondylar ridge of femur and adjacent part of its popliteal surface. Oblique popliteal ligament of knee joint.

Insertion
Posterior surface of calcaneus (or sometimes into the medial surface of the tendo calcaneus).

Action
Plantar flexes ankle joint. Feebly flexes knee joint.

Nerve
Tibial nerve, L4, **5**, S1, (2).

Artery
Sural branches of the popliteal artery (a continuation of the femoral artery).

Basic functional movement
Example: Standing on tip-toes.

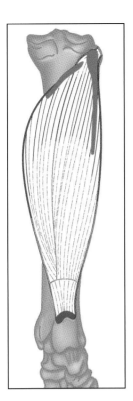

Posterior view,
right leg.

Latin, sole-shaped (fish).

Part of the triceps surae. The soleus is so called because its shape resembles a fish. The calcaneal tendon of the soleus and gastrocnemius is the thickest and strongest tendon in the body.

Origin
Posterior surfaces of head of fibula and upper third of body of fibula. Soleal line and middle third of medial border of tibia. Tendinous arch between tibia and fibula.

Insertion
With tendon of gastrocnemius into posterior surface of calcaneus.

Action
Plantar flexes ankle joint. The soleus is frequently in contraction during standing to prevent the body falling forwards at the ankle joint; i.e. to offset the line of pull through the body's centre of gravity. Thus, it helps to maintain the upright posture.

Nerve
Tibial nerve, L5, S1, 2.

Artery
Posterior tibial artery
(from popliteal artery).
Sural branches of the popliteal artery, and the peroneal artery
via posterior tibial artery.

Basic functional movement
Example: Standing on tip-toes.

POPLITEUS

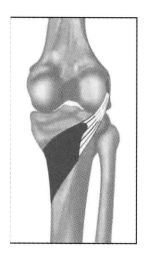

Posterior view, right leg.

Latin, *poples,* ham.

The tendon from the origin of popliteus lies inside the capsule of the knee joint.

Origin
Lateral surface of lateral condyle of femur. Oblique popliteal ligament of knee joint.

Insertion
Upper part of posterior surface of tibia, superior to soleal line.

Action
Laterally rotates femur on tibia when foot is fixed on the ground. Medially rotates tibia on femur when the leg is non-weight bearing. Assists flexion of knee joint, (popliteus 'unlocks' the extended knee joint to initiate flexion of the leg). Helps reinforce posterior ligaments of knee joint.

Nerve
Tibial nerve, L4, 5, S1.

Artery
Sural and medial inferior genicular branches of popliteal artery
(a continuation of the femoral artery).

Basic functional movement
Example: Walking.

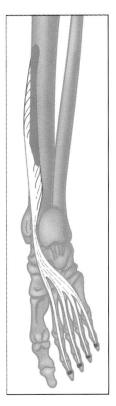

Posterior view,
right leg.

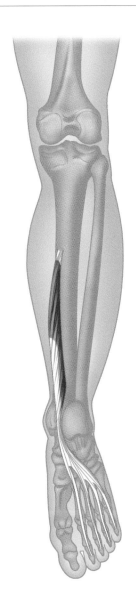

Latin, *flex*, to bend; *digit*, toe; *longus*, long.

The insertion of the tendons of this muscle into the lateral four toes parallels the insertion of flexor digitorum profundus in the hand.

Origin
Medial part of posterior surface of tibia, below soleal line.

Insertion
Bases of distal phalanges of second through fifth toes.

Action
Flexes all the joints of the lateral four toes (enabling the foot to firmly grip the ground when walking). Helps to plantar flex the ankle joint and invert the foot.

Nerve
Tibial nerve, L5, S1, (2).

Artery
Posterior tibial artery
(from popliteal artery).

Basic functional movement
Examples: Walking (esp. bare foot on uneven ground). Standing on tip-toes.

FLEXOR HALLUCIS LONGUS

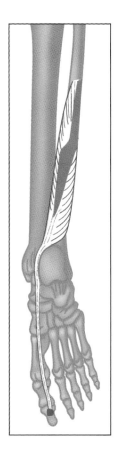

Posterior view,
right leg.

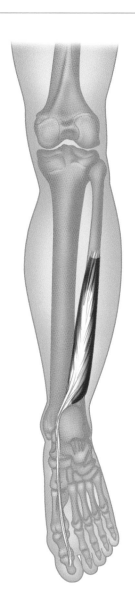

Latin, *flex,* to bend; *hallux,* great toe; *longus,* long.

This muscle helps maintain the medial longitudinal arch of the foot.

Origin
Lower two-thirds of posterior surface of fibula. Interosseous membrane. Adjacent intermuscular septum.

Insertion
Base of distal phalanx of great toe.

Action
Flexes all the joints of the great toe, and is important in the final propulsive thrust of the foot during walking. Helps to plantar flex the ankle joint and invert the foot.

Nerve
Tibial nerve, L**5**, S**1**, **2.**

Artery
Peroneal (fibular) artery
via posterior tibial artery (from popliteal artery).

Basic functional movement
Examples: Pushing off the surface in walking (esp. bare foot on uneven ground). Standing on tip-toes.

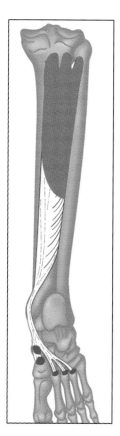

Posterior view,
right leg.

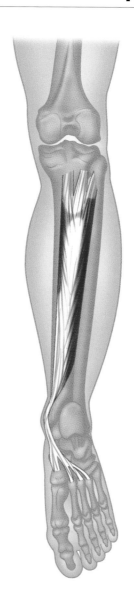

Latin, *tibia,* pipe or flute / shinbone; *posterior,* behind.

Tibialis posterior is the deepest muscle on the back of the leg. It helps maintain the arches of the foot.

Origin
Lateral part of posterior surface of tibia. Upper two-thirds of posterior surface of fibula. Most of interosseous membrane.

Insertion
Tuberosity of navicular. By fibrous expansions to the sustentaculum tali, three cuneiforms, cuboid and bases of the second, third and fourth metatarsals.

Action
Inverts the foot. Assists in plantar flexion of the ankle joint.

Nerve
Tibial nerve, L(4), **5, S1.**

Artery
Peroneal (fibular) artery
via posterior tibial artery (from popliteal artery).

Basic functional movement
Examples: Standing on tip-toes. Pushing down car pedals.

MUSCLES OF THE FOOT

There are four layers of muscle in the sole of the foot. The **first layer** is the most inferior (that is, the most superficial and closest to the ground in standing), comprising *abductor hallucis, flexor digitorum brevis,* and *abductor digiti minimi*. The **second layer** contains the *lumbricales* and *quadratus plantae*, plus the tendons of flexor hallucis longus and flexor digitorum longus. The **third layer** contains *flexor hallucis brevis, adductor hallucis,* and *flexor digiti minimi brevis*. The uppermost **fourth layer** contains the *interossei* and the tendons of tibialis posterior and peroneus longus. On the dorsum of the foot lies *extensor digitorum brevis*.

Abductor hallucis

Abductor digiti minimi

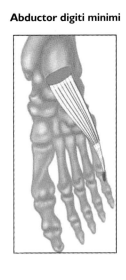

Lumbricales

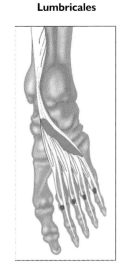

Adductor hallucis

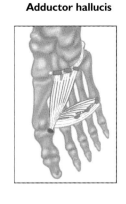

Plantar interossei

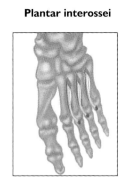

Flexor digitorum brevis

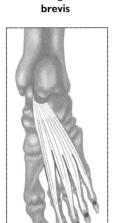

Quadratus plantae

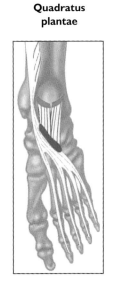

Flexor hallucis brevis

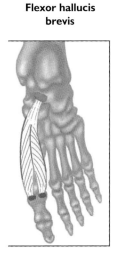

Extensor digitorum brevis

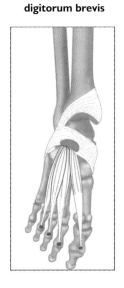

Dorsal interossei

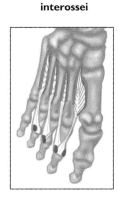

Flexor digiti minimi brevis

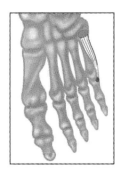

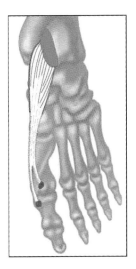

Plantar view, right foot.

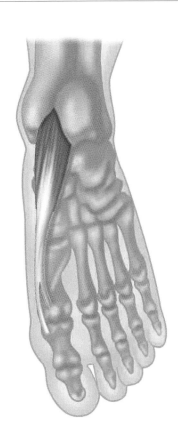

Latin, *abduct*, away from; *hallux*, great toe.

First layer: most superficial layer of muscles of sole of foot. Includes: *abductor hallucis, flexor digitorum brevis*, and *abductor digiti minimi*. Abductor hallucis forms the medial margin of the sole of the foot.

Origin
Tuberosity of calcaneus. Flexor retinaculum. Plantar aponeurosis.

Insertion
Medial side of base of proximal phalanx of great toe.

Action
Abducts and helps flex great toe at metatarsophalangeal joint.

Nerve
Medial plantar nerve, L4, **5**, **S1**.

Artery
Medial plantar artery
(from posterior tibial artery).

Basic functional movement
Helps foot stability and power in walking and running.

FLEXOR DIGITORUM BREVIS

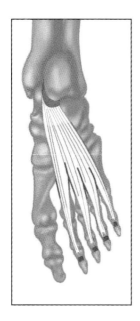

Plantar view, right foot.

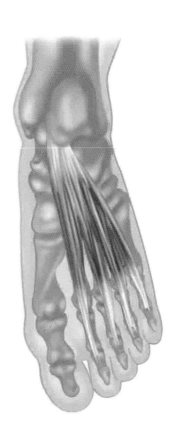

Latin, *flex*, to bend; *digit*, toe; *brevis*, short.

First layer: most superficial layer of muscles of sole of foot. Includes: *abductor hallucis, flexor digitorum brevis, abductor digiti minimi*. Flexor digitorum brevis is equivalent to the flexor digitorum superficialis muscle of the arm.

Origin
Tuberosity of calcaneus. Plantar aponeurosis. Adjacent intermuscular septa.

Insertion
Middle phalanges of second to fifth toes.

Action
Flexes all the joints of the lateral four toes except the distal interphalangeal joints.

Nerve
Medial plantar nerve, L4, **5**, S1.

Artery
Medial plantar artery
(from posterior tibial artery).

Basic functional movement
Helps foot stability and power in walking and running.

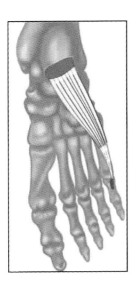

Plantar view, right foot.

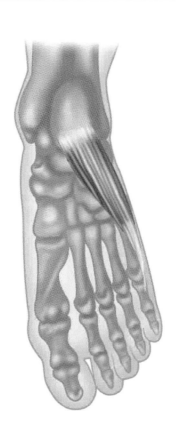

Latin, *abduct*, away from; *digit*, toe; *minimi*, smallest.

First layer: most superficial layer of muscles of sole of foot. Includes: *abductor hallucis, flexor digitorum brevis, abductor digiti minimi*. Abductor digiti minimi forms the lateral margin of the sole of the foot.

Origin
Tuberosity of calcaneus. Plantar aponeurosis. Adjacent intermuscular septa.

Insertion
Lateral side of base of proximal phalanx of fifth toe.

Action
Abducts fifth toe.

Nerve
Lateral plantar nerve, S**2**, 3.

Artery
Lateral plantar artery
(from posterior tibial artery).

QUADRATUS PLANTAE

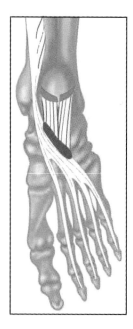

Plantar view, right foot.

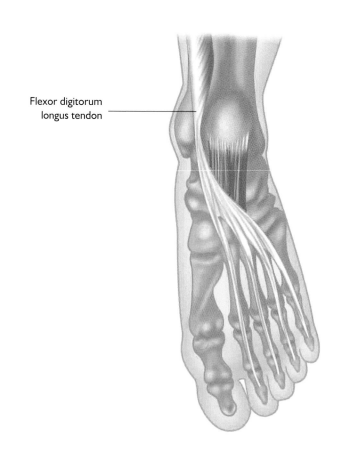

Flexor digitorum
longus tendon

Latin, *quadratus*, squared; *planta*, sole of the foot.

Second layer: includes: *quadratus plantae* (also called *flexor accessorius*), and *lumbricales*. Quadratus plantae has no counterpart in the hand.

Origin
Medial head: medial surface of calcaneus.
Lateral head: lateral border of inferior surface of calcaneus.

Insertion
Lateral border of tendon of flexor digitorum longus.

Action
Flexes distal phalanges of second through to fifth toes. Modifies the oblique line of pull of the flexor digitorum longus tendons to bring it in line with the long axis of the foot.

Nerve
Lateral plantar nerve, S**1**, **2**.

Artery
Lateral plantar artery
(from posterior tibial artery).

Basic functional movement
Example: Holding a pencil between the toes and the ball of the foot.

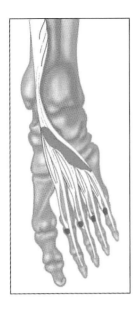

Plantar view, right foot.

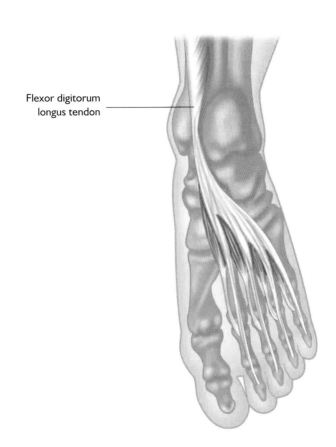

Flexor digitorum longus tendon

Latin, earthworm.

Second layer: includes: *quadratus plantae* and the four *lumbricales*.

Origin
Tendons of flexor digitorum longus.

Insertion
Medial side of base of proximal phalanges of second through to fifth toes and corresponding extensor expansion.

Action
Flex the metatarsophalangeal joints and extend the interphalangeal joints of the lateral four toes.

Nerve
Lateral three lumbricales: lateral plantar nerve L (4), (5), S**1**, **2**.
First lumbricalis: medial plantar nerve, L4, **5**, S1.

Artery
**1st: medial plantar artery, 2nd–4th: lateral plantar artery
(from posterior tibial artery).**

Basic functional movement
Example: Gathering up material under the foot using the toes only.

FLEXOR HALLUCIS BREVIS

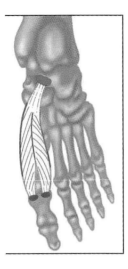

Plantar view, right foot.

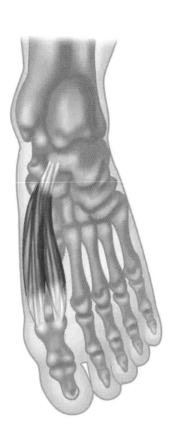

Latin, *flex*, to bend; *hallux*, great toe; *brevis*, short.

Third layer: includes *flexor hallucis brevis, adductor hallucis,* and *flexor digiti minimi brevis*. The tendons of flexor hallucis brevis contain sesamoid bones. During walking, the great toe pivots on these sesamoid bones.

Origin
Medial part of plantar surface of cuboid bone. Adjacent part of lateral cuneiform bone. Tendon of tibialis posterior.

Insertion
Medial part: medial side of base of proximal phalanx of great toe.
Lateral part: lateral side of base of proximal phalanx of great toe.

Action
Flexes the metatarsophalangeal joint of the great toe.

Nerve
Medial plantar nerve, L4, **5**, S**1**.

Artery
Medial plantar artery
(from posterior tibial artery).

Basic functional movement
Example: Helping to gather up material under the foot by involving the big toe.

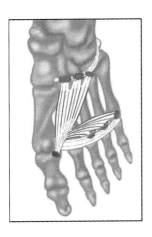

Plantar view, right foot.

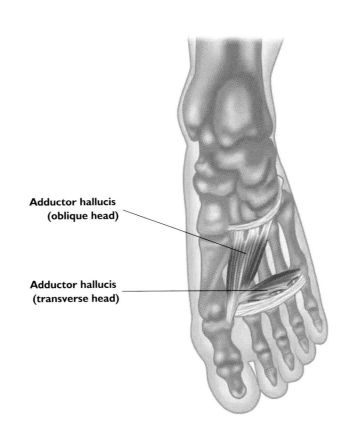

Adductor hallucis
(oblique head)

Adductor hallucis
(transverse head)

Latin, *adduct*, toward; *hallux*, great toe.

Third layer: includes *flexor hallucis brevis, adductor hallucis, flexor digiti minimi brevis*. Similar to the adductor of the thumb, the adductor hallucis has two heads.

Origin
Oblique head: bases of second, third and fourth metatarsals. Sheath of peroneus longus tendon.
Transverse head: plantar metatarsophalangeal ligaments of third, fourth and fifth toes. Transverse metatarsal ligaments.

Insertion
Lateral side of base of proximal phalanx of great toe.

Action
Adducts and assists in flexing the metatarsophalangeal joint of the great toe.

Nerve
Lateral plantar nerve, S1, 2.

Artery
Medial plantar artery
(from posterior tibial artery).

Basic functional movement
Example: Making a space between the big toe and the adjacent toe.

FLEXOR DIGITI MINIMI BREVIS

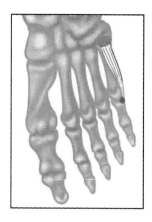

Plantar view, right foot.

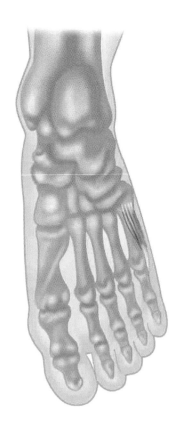

Latin, *flex*, to bend; *digit*, toe; *minimi*, smallest; *brevis*, short.

Third layer: includes *flexor hallucis brevis, adductor hallucis, flexor digiti minimi brevis*.

Origin
Sheath of peroneus longus tendon. Base of fifth metatarsal.

Insertion
Lateral side of base of proximal phalanx of little toe.

Action
Flexes the little toe at the metatarsophalangeal joint.

Nerve
Lateral plantar nerve, S**2**, 3.

Artery
Lateral plantar artery
(from posterior tibial artery).

Basic functional movement
Example: Works alongside other toes to gather up material under the foot.

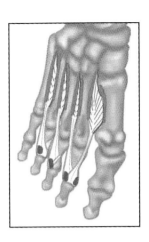

Dorsal view, right foot.

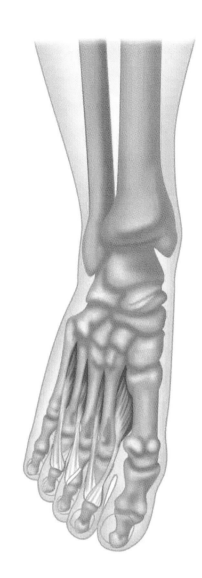

Latin, *dorsum*, back; *interosseus*, between bones.

Fourth layer: deepest (most superior) layer of muscles of sole of foot. Consists of the four muscles of the *dorsal interossei* and the three muscles of the *plantar interossei*. Similar to the hand, the dorsal interossei are larger than the plantar interossei.

Origin
Adjacent sides of metatarsal bones.

Insertion
Bases of proximal phalanges:
First: medial side of proximal phalanx of second toe.
Second to fourth: lateral sides of proximal phalanges of second to fourth toes.

Action
Abduct (spreads) toes. Flex metatarsophalangeal joints.

Nerve
Lateral plantar nerve, S1, 2.

Artery
Dorsal metatarsal arteries
via arcuate artery of foot, (from dorsalis pedis artery, a continuation of anterior tibial artery).

Basic functional movement
Example: Facilitates walking.

PLANTAR INTEROSSEI

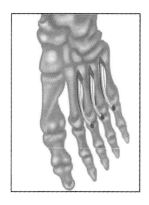

Plantar view, right foot.

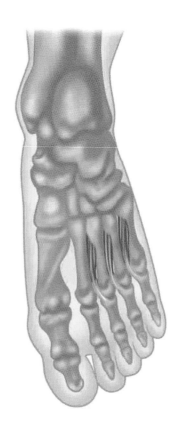

Latin, *planta*, sole of the foot; *interosseus*, between bones.

Fourth layer: deepest layer of muscles of sole of foot.

Origin
Bases and medial sides of third, fourth and fifth metatarsals.

Insertion
Medial sides of bases of proximal phalanges of same toes.

Action
Adduct (close together) toes. Flex metatarsophalangeal joints.

Nerve
Lateral plantar nerve, S**1**, **2**.

Artery
Plantar metatarsal arteries
via plantar arch of lateral plantar artery (from posterior tibial artery).

Basic functional movement
Example: Facilitates walking.

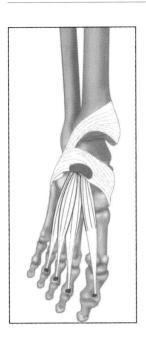

Dorsal view, right foot.

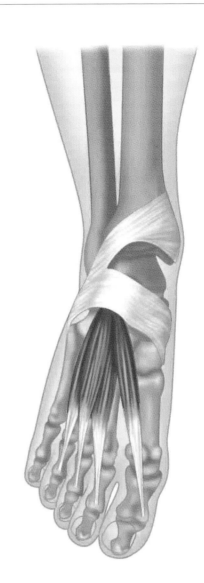

Latin, *extensor*, to extend; *digit*, toe; *brevis*, short.

This is the only muscle arising from the dorsum of the foot. The part of the extensor digitorum brevis that goes to the great toe is called the *extensor hallucis brevis*.

Origin
Anterior part of superior and lateral surfaces of calcaneus. Lateral talocalcaneal ligament. Inferior extensor retinaculum.

Insertion
Base of proximal phalanx of great toe. Lateral sides of tendons of extensor digitorum longus to second, third and fourth toes.

Action
Extends the joints of the medial four toes.

Nerve
Deep fibular (peroneal) nerve, L4, **5**, S1.

Artery
Dorsalis pedis artery
(a continuation of anterior tibial artery).

Basic functional movement
Example: Facilitates walking.

Main Muscles Involved in Different Movements of the Body

	MOVEMENT	MUSCLE
MANDIBLE	*Elevation*	Temporalis (anterior fibres); Masseter; Pterygoideus Medialis
	Depression	Pterygoideus Lateralis; Digastricus; Mylohyoideus; Geniohyoideus
	Protraction	Pterygoideus Lateralis; Pterygoideus Medialis; Masseter (superficial fibres)
	Retraction	Temporalis (horizontal fibres); Digastricus
	Chewing	Pterygoideus Lateralis; Pterygoideus Medialis; Masseter; Temporalis
LARYNX	*Elevation*	Digastricus; Stylohyoideus; Mylohyoideus; Geniohyoideus; Thyrohyoideus
	Depression	Sternohyoideus; Sternothyroideus; Omohyoideus
	Protraction	Geniohyoideus
	Retraction	Stylohyoideus
ATLANTO-OCCIPITAL & ATLANTO-AXIAL JOINTS	*Flexion*	Longus Capitis; Rectus Capitis Anterior; Sternocleidomastoideus (anterior fibres)
	Extension	Semispinalis Capitis; Splenius Capitis; Rectus Capitis Posterior Major; Rectus Capitis Posterior Minor; Obliquus Capitis Superior; Longissimus Capitis; Trapezius; Sternocleidomastoideus (posterior fibres)
	Rotation and Lateral Flexion	Sternocleidomastoideus; Obliquus Capitis Inferior; Obliquus Capitis Superior; Rectus Capitis Lateralis; Longissimus Capitis; Splenius Capitis
INTERVERTEBRAL JOINTS CERVICAL REGION	*Flexion*	Longus Colli; Longus Capitis; Sternocleidomastoideus
	Extension	Longissimus Cervicis; Longissimus Capitis; Splenius Capitis; Splenius Cervicis; Semispinalis Cervicis; Semispinalis Capitis; Trapezius; Interspinales; Iliocostalis Cervicis
	Rotation and Lateral Flexion	Longissimus Cervicis; Longissimus Capitis; Splenius Capitis; Splenius Cervicis; Multifidis; Longus Colli; Scalenus Anterior; Scalenus Medius; Scalenus Posterior; Sternocleidomastoideus; Levator Scapulae; Iliocostalis Cervicis; Intertransversarii
INTERVERTEBRAL JOINTS THORACIC/LUMBAR REGIONS	*Flexion*	Muscles of Anterior Abdominal Wall
	Extension	Erector Spinae; Quadratus Lumborum; Trapezius
	Rotation and Lateral Flexion	Iliocostalis Lumborum; Iliocostalis Thoracis; Multifidis; Rotatores; Intertransversarii; Quadratus Iliocostalis Lumborum; Iliocostalis Thoracis; Multifidis; Rotatores; Intertransversarii; Quadratus Lumborum; Psoas Major; Muscles of Anterior Abdominal Wall

	MOVEMENT	MUSCLE
SHOULDER GIRDLE	Elevation	Trapezius (upper fibres); Levator Scapulae; Rhomboideus Minor; Rhomboideus Major; Sternocleidomastoideus
	Depression	Trapezius (lower fibres); Pectoralis Minor; Pectoralis Major (sternocostal portion); Latissimus Dorsi
	Protraction	Serratus Anterior; Pectoralis Minor; Pectoralis Major
	Retraction	Trapezius (middle fibres); Rhomboideus Minor; Rhomboideus Major; Latissimus Dorsi
	Lateral Displacement of Inferior Angle of Scapula	Serratus Anterior; Trapezius (upper and lower fibres)
	Medial Displacement of Inferior Angle of Scapula	Pectoralis Minor; Rhomboideus Minor; Rhomboideus Major; Latissimus Dorsi
SHOULDER JOINT	Flexion	Deltoideus (anterior portion); Pectoralis Major (clavicular portion : sternocostal portion flexes the extended humerus as far as the position of rest); Biceps Brachii; Coracobrachialis
	Extension	Deltoideus (posterior portion); Teres Major (of flexed humerus); Latissimus Dorsi (of flexed humerus); Pectoralis Major (sternocostal portion of flexed humerus); Triceps Brachii (long head to position of rest)
	Abduction	Deltoideus (middle portion); Supraspinatus; Biceps Brachii (long head)
	Adduction	Pectoralis Major; Teres Major; Latissimus Dorsi; Triceps Brachii (long head); Coracobrachialis
	Lateral Rotation	Deltoideus (posterior portion); Infraspinatus; Teres Minor
	Medial Rotation	Pectoralis Major; Teres Major; Latissimus Dorsi; Deltoideus (anterior portion); Subscapularis
	Horizontal Flexion	Deltoideus (anterior portion); Pectoralis Major; Subscapularis
	Horizontal Extension	Deltoideus (posterior portion); Infraspinatus
ELBOW JOINT	Flexion	Brachialis; Biceps Brachii; Brachioradialis; Extensor Carpi Radialis Longus; Pronator Teres; Flexor Carpi Radialis
	Extension	Triceps Brachii; Anconeus

	MOVEMENT	MUSCLE
RADIO-ULNAR JOINTS	*Supination*	Supinator; Biceps Brachii; Extensor Pollicis Longus
	Pronation	Pronator Quadratus; Pronator Teres; Flexor Carpi Radialis
RADIOCARPAL AND MIDCARPAL JOINTS	*Flexion*	Flexor Carpi Radialis; Flexor Carpi Ulnaris; Palmaris Longus; Flexor Digitorum Superficialis; Flexor Digitorum Profundus; Flexor Pollicis Longus; Abductor Pollicis Longus; Extensor Pollicis Brevis
	Extension	Extensor Carpi Radialis Brevis; Extensor Carpi Radialis Longus; Extensor Carpi Ulnaris; Extensor Digitorum; Extensor Indicis; Extensor Pollicis Longus; Extensor Digiti Minimi
	Abduction	Extensor Carpi Radialis Brevis; Extensor Carpi Radialis Longus; Flexor Carpi Radialis; Abductor Pollicis Longus; Extensor Pollicis Longus; Extensor Pollicis Brevis
	Adduction	Flexor Carpi Ulnaris; Extensor Carpi Ulnaris
METACARPOPHALANGEAL JOINTS OF THE FINGERS	*Flexion*	Flexor Digitorum Profundus; Flexor Digitorum Superficialis; Lumbricales; Interossei; Flexor Digiti Minimi; Abductor Digiti Minimi; Palmaris Longus (through palmar aponeurosis)
	Extension	Extensor Digitorum; Extensor Indicis; Extensor Digiti Minimi
	Abduction and Adduction	Interossei; Abductor Digiti Minimi; Lumbricales (may assist in radial deviation); Extensor Digitorum (abducts by hyperextending; tendon to index radially deviates); Flexor Digitorum Profundus (adducts by flexing); Flexor Digitorum Superficialis (adducts by flexing)
	Rotation	Lumbricales; Interossei (movement slight except index; only effective when phalanx is flexed); Opponens Digiti Minimi (rotates little finger at carpometacarpal joint)
INTERPHALANGEAL JOINTS OF THE FINGERS	*Flexion*	Flexor Digitorum Profundus (both joints); Flexor Digitorum Superficialis (proximal joint only)
	Extension	Extensor Digitorum; Extensor Digiti Minimi; Extensor Indicis; Lumbricales; Interossei

	MOVEMENT	MUSCLE
CARPOMETACARPAL JOINT OF THE THUMB	*Flexion*	Flexor Pollicis Brevis; Flexor Pollicis Longus; Opponens Pollicis
	Extension	Extensor Pollicis Brevis; Extensor Pollicis Longus; Abductor Pollicis Longus
	Abduction	Abductor Pollicis Brevis; Abductor Pollicis Longus
	Adduction	Adductor Pollicis; Dorsal Interossei (first only); Extensor Pollicis Longus (in full extension / abduction); Flexor Pollicis Longus (in full extension / abduction)
	Opposition	Opponens Pollicis; Abductor Pollicis Brevis; Flexor Pollicis Brevis; Flexor Pollicis Longus; Adductor Pollicis
METACARPOPHALANGEAL JOINT OF THE THUMB	*Flexion*	Flexor Pollicis Brevis; Flexor Pollicis Longus; Palmar Interossei (first only); Abductor Pollicis Brevis
	Extension	Extensor Pollicis Brevis; Extensor Pollicis Longus
	Abduction	Abductor Pollicis Brevis
	Adduction	Adductor Pollicis; Palmar Interossei (first only)
INTERPHALANGEAL JOINT OF THE THUMB	*Flexion*	Flexor Pollicis Longus
	Extension	Abductor Pollicis Brevis; Extensor Pollicis Longus; Adductor Pollicis; Extensor Pollicis Brevis (occasional insertion)
HIP JOINT	*Flexion*	Iliopsoas; Rectus Femoris; Tensor Fasciae Latae; Sartorius; Adductor Brevis; Adductor Longus; Pectineus
	Extension	Gluteus Maximus; Semitendinosus; Semimembranosus; Biceps Femoris (long head); Adductor Magnus (ischial fibres)
	Abduction	Gluteus Medius; Gluteus Minimus; Tensor Fasciae Latae; Obturator Internus (in flexion); Piriformis (in flexion)
	Adduction	Adductor Magnus; Adductor Brevis; Adductor Longus; Pectineus; Gracilis; Gluteus Maximus (lower fibres); Quadratus Femoris
	Lateral Rotation	Gluteus Maximus; Obturator Internus; Gemelli; Obturator Externus; Quadratus Femoris; Piriformis; Sartorius; Adductor Magnus; Adductor Brevis; Adductor Longus
	Medial Rotation	Iliopsoas (in initial stage of flexion); Tensor Fasciae Latae; Gluteus Medius (anterior fibres); Gluteus Minimus (anterior fibres)

	MOVEMENT	MUSCLE
KNEE JOINT	*Flexion*	Semitendinosus; Semimembranosus; Biceps Femoris; Gastrocnemius; Plantaris; Sartorius; Gracilis; Popliteus
	Extension	Quadratus Femoris
	Medial Rotation of Tibia on Femur	Popliteus; Semitendinosus; Semimembranosus; Sartorius; Gracilis
	Lateral Rotation of Tibia on Femur	Biceps Femoris
ANKLE JOINT	*Dorsiflexion*	Tibialis Anterior, Extensor Hallucis Longus; Extensor Digitorum Longus; Fibularis (Peroneus) Tertius
	Plantar Flexion	Gastrocnemius; Plantaris; Soleus; Tibialis Posterior; Flexor Hallucis Longus; Flexor Digitorum Longus; Fibularis (Peroneus) Longus; Fibularis (Peroneus) Brevis
INTERTARSAL JOINTS	*Inversion*	Tibialis Anterior; Tibialis Posterior
	Eversion	Fibularis (Peroneus) Tertius; Fibularis (Peroneus) Longus; Fibularis (Peroneus) Brevis
	Other Movements	Sliding movements which allow some dorsiflexion, plantar flexion, abduction and adduction, are produced by the muscles acting on the toes. Tibialis Anterior, Tibialis Posterior, and Fibularis (Peroneus) Tertius are also involved.
METATARSOPHALANGEAL JOINTS OF THE TOES	*Flexion*	Flexor Hallucis Brevis; Flexor Hallucis Longus; Flexor Digitorum Longus; Flexor Digitorum Brevis; Flexor Digiti Minimi Brevis; Lumbricales; Interossei
	Extension	Extensor Hallucis Longus; Extensor Digitorum Brevis; Extensor Digitorum Longus
	Abduction and Adduction	Abductor Hallucis; Adductor Hallucis; Interossei; Abductor Digiti Minimi
INTERPHALANGEAL JOINTS OF THE TOES	*Flexion*	Flexor Hallucis Longus; Flexor Digitorum Brevis (proximal joint only); Flexor Digitorum Longus
	Extension	Extensor Hallucis Longus; Extensor Digitorum Brevis (not in great toe); Extensor Digitorum Longus; Lumbricales

Resources

Alter, M. J.: 1998. *Sport Stretch: 311 Stretches for 41 Sports*. Human Kinetics, Champaign.

Anderson, D. M. (chief Lexicographer): 2003. *Dorland's Illustrated Medical Dictionary, 30th edition*. Saunders, an imprint of Elsevier, Philadelphia.

Biel, A.: 2001. *Trail Guide to the Body, 2nd edition*. Books of Discovery, Boulder.

Clemente, C. M. (editor): 1985. *Gray's Anatomy of the Human Body, 30th edition*. Lea & Febiger, Philadelphia.

DeJong, R. N.: 1967. *The Neurological Examination, 2nd & 3rd editions*. Harper & Row, New York.

Department of Anatomy, University of Arkansas for Medical Science. 1999. *University Website / Database*.

Foerster, O., and Bumke, O.: 1936. *Handbuch der Neurologie (vol. V)*. Publisher unknown, Breslau.

Haymaker, W., and Woodhall, B.: 1953. *Peripheral Nerve Injuries, 2nd edition*. W. B. Saunders Co., Philadelphia.

Kendall, F. P., and McCreary, E. K.: 1983. *Muscles, Testing & Function, 3rd edition*. Williams & Wilkins, Baltimore.

Lawrence, M.: 2004. *Complete Guide to Core Stability*. A & C Black, London.

Norris, C. M.: 1997. *Abdominal Training*. A & C Black, London.

Romanes, G. J. (editor): 1972. *Cunningham's Textbook of Anatomy, 11th edition*. Oxford University Press, London.

Schade, J. P.: 1966. *The Peripheral Nervous System*. Elsevier, New York.

Spalteholz, W.: (date unknown). *Hand Atlas of Human Anatomy (vols. II & III, 6th edition)*. J. B. Lippincott, London.

Tortora, G.: 1989. *Principles of Human Anatomy, 5th edition*. Harper & Row, New York.

General Index

Index of Muscles

CHANNELS AND ACU-POINTS

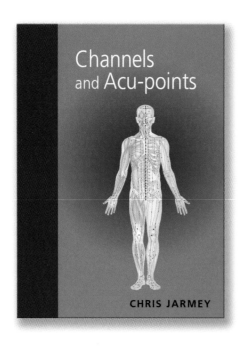

AUTHOR:	Chris Jarmey	FORMAT:	275 mm x 212 mm
ISBN:	0 9543188 4 6 (UK)	COVER:	Full colour
ISBN:	1 55643 535 5 (US)	ILLUSTRATIONS:	240 full colour
PRICE:	£19.99		illustrations
PRICE:	$39.95	BINDING:	Paperback
PAGES:	384		

Channels and Acu-points is an accurate reference for the location of points and of the primary channels of energy (Qi) within the human body. Students and practitioners of acupuncture and oriental bodywork will appreciate the ease-of-use attributes of this book, such as the colour coding of points according to their category, and the overall clarity of presentation. Deep channel pathways and musculo-cutaneous (sinew) branches for each channel are included.

THE CONCISE BOOK OF MUSCLES

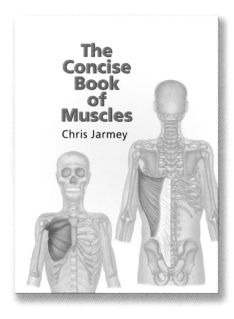

AUTHOR:	Chris Jarmey	FORMAT:	275 mm x 212 mm
ISBN:	0 9543188 1 1 (UK)	COVER:	Full colour
ISBN:	1 55643 466 9 (US)	ILLUSTRATIONS:	140 full colour illustrations
PRICE:	£16.99		110 line drawings
PRICE:	$29.95	BINDING:	Paperback
PAGES:	160		

The Concise Book of Muscles has been written for the student and early practitioner of anatomy and physiology, massage / bodywork, physical therapy, chiropractic, medicine, physiotherapy, or any other health-related field. Containing full colour illustrations, the book is a compact reference guide, clearly identifying all the major muscles, showing the origin, insertion, action and innervation of each muscle.

The Concise Book of Muscles also uniquely illustrates examples of stretching and strengthening exercises for each muscle, which allows the reader to develop an understanding of the mechanics of movement.

Amanda Williams (Illustrator) graduated from Middlesex University with a first-class honours degree in Graphic Design (Scientific Illustration). Since then she has worked as a freelance illustrator for various publishers, including Harcourt Brace, Elsevier Science, and The Royal College of Surgeons.

Chris Jarmey, M.C.S.P., D.S., M.R.S.S., qualified as a Chartered Physiotherapist in 1979. He also studied acupuncture and osteopathy in the early 1980s. Chris teaches body mechanics, bodywork therapy, and anatomy extensively throughout Europe.